New Perspectives in Clinical Microbiology

Series editor: W. Brumfitt

Volume 2

New Perspectives in Clinical Microbiology

Series editor: W. Brumfitt

Volume 1. New Perspectives in Clinical Microbiology, edited by W. Brumfitt.
Assistant-editor: J. M. T. Hamilton-Miller.
Kluwer Medical - London/Martinus Nijhoff - The Hague 1978. ISBN 90 247 2074 5

Volume 2. Aspects of Slow and Persistent Virus Infections, edited by D. A. J. Tyrrell.
Martinus Nijhoff - The Hague - Boston - London 1979. ISBN 90 247 2281 0

Volume 3. Combined Antimicrobial Therapy, edited by W. Brumfitt, L. Curcio and
L. Silvestri. Martinus Nijhoff - The Hague - Boston - London 1979. ISBN 90 247 2280 2

Aspects of Slow and Persistent Virus Infections

Proceedings of the European Workshop
sponsored by the Commission of the European
Communities on the advice of the Committee on
Medical and Public Health Research, held in
London (U.K.), April 5-6, 1979

Edited by

D. A. J. Tyrrell

Clinical Research Centre, Harrow

Martinus Nijhoff Publishers - The Hague/Boston/London 1979
for
The Commission of the European Communities

The distribution of this book is handled by the following team of publishers:

for the United States and Canada

Kluwer Boston, Inc.
160 Old Derby Street
Hingham, MA 02043
USA

for all other countries

Kluwer Academic Publishers Group
Distribution Center
P.O. Box 322
3300 AH Dordrecht
The Netherlands

Library of Congress Cataloging in Publication Data   CIP

Main entry under title:

Aspects of Slow and Persistent Virus Infections.

(New Perspectives in Clinical Microbiology = V. 2) + Papers and Discussions at a Workshop . . . in London on the 5th and 6th of April 1979 as a Part of the Programme of the the Commission of the European Communities on Medical and Public Health Research.
   Includes index.
   1.   Virus Diseases, Slow—Congresses.   I.   Tyrrell, David Arthur John.
II.   Commission of the European Communities.
RC114.6.A84   616.9'2   79-26296
ISBN-13:978-94-009-9341-9   e-ISBN-13:978-94-009-9339-6
DOI: 10.1007/978-94-009-9339-6

ISBN-13:978-94-009-9341-9

Publication arranged by
Commission of the European Communities,
Directorate-General for Scientific and Technical Information and Information Management
Luxembourg

EUR 6582 EN

LEGAL NOTICE
Neither the Commission of the European Communities or any person acting on behalf of the Commission is responsible for the use which might be made of the following information.

For further information:
Martinus Nijhoff Publishers b.v.,
P.O. Box 566
2501 CN The Hague
The Netherlands

PREFACE

     This book records the papers and discussions at a Workshop which took place in London on the 5th and 6th of April 1979, as part of the programme of the Commission of the European Communities on Medical and Public Health Research. However the views expressed are those of the individuals concerned and not of the EEC or any of its organs. The object was to discuss certain biological aspects of natural and experimental slow virus infections. Because the amount of knowledge and the focus of interest varied in respect of each infection the approach and emphasis varied also. In the case of scrapie, we discussed the nature of the agent and the mode of pathogenesis, in the case of SSPE, the search for unusual features of the virus, and recent detailed work on the immunology of the disease. As for Visna we reviewed the present understanding of the virus and its pathogenicity and also field epidemiology and methods for its control. There were also general papers, on interferon and oncornaviruses for example. We thank all those who made the meeting possible and enabled us to produce this book quickly, so that those who could not attend the meeting may nevertheless be able to read a great deal of what went on at it. In particular we would thank the Ciba Foundation who allowed us the use of their premises and Mrs. Jean Ashley who dealt with most of the arrangements. Last but not least we thank Dr. R.N.P. Sutton who as supporter and discussion editor rapidly produced a summary of the discussion which took place.

D.A.J. Tyrrell

TABLE OF CONTENTS

LIST OF CONTRIBUTORS

G. Agnarsdottir
The Royal Postgraduate Medical School
Departments of Immunology and Virology
Du Cane Road
LONDON, W 12
England

M. Bergeret
Institut National de la Santé et de la Recherche Médicale
u.-43, Hôpital Saint-Vincent-de-Paul
PARIS, 75014
France

G.F. de Boer
Centraal Diergeneeskundig Instituut
Afd. Virologie
Houtribweg 39
8221 RA LELYSTAD
The Netherlands

M.F. Bourgade
Institut National de la Santé et de la Recherche Médicale
u.-43, Hôpital Saint-Vincent-de-Paul
PARIS, 75014
France

C. Chany
Institut National de la Santé et de la Recherche Médicale
u.-43, Hôpital Saint-Vincent-de-Paul
PARIS, 75014
France

F. Chany-Fournier
Institut National de la Santé et de la Recherche Médicale
u.-43, Hôpital Saint-Vincent-de-Paul
PARIS, 75014
France

P.W. Ewan
The Royal Postgraduate Medical School
Departments of Immunology and Virology
Du Cane Road
LONDON, W 12
England

H. Fraser
ARC Animal Breeding Research Organisation
King's Buildings, West Mains Road
EDINBURGH, EH9 3JQ
Scotland

K.B. Fraser
Department of Microbiology and Immunology
The Queen's University of Belfast
Grosvenor Road
BELFAST, BT12 6BN
Northern Ireland

G. Georgsson
Institute for Experimental Pathology
University of Iceland
KELDUR,REYKJAVIK
Iceland

N.T. Gorman
MRC Group on Mechanisms in Tumour Immunity
The Medical School
Hills Road
CAMBRIDGE
England

J. Habicht
MRC Group on Mechanisms in Tumour Immunity
The Medical School
Hills Road
CAMBRIDGE
England

D.J. Houwers
Centraal Diergeneeskundig Instituut
Afd. Virologie
Houtribweg 39
8221 RA LELYSTAD
The Netherlands

R.H. Kimberlin
Institute for Research on Animal Diseases
Compton, Near Newbury
BERKS, RG16 ONN
England

W. Kreth
Institut für Virologie und Immunologie
der Universitat Wurzburg
Versbacher Landstrasse 7
WURZBURG 8700
Germany

P. Lachmann
Addenbrookes Hospital
Hills Road
CAMBRIDGE, CB2 2QQ
England

E. Lund
Department of Veterinary Virology and Immunology
The Royal Veterinary and Agricultural
University of Copenhagen
13, Bulowsvej
1870 COPENHAGEN
Denmark

J.R. Martin
Institute for Experimental Pathology
University of Iceland
Keldur
REYKJAVÍK
Iceland

V. ter Meulen
Institut für Virologie und Immunbiologie
der Universtität Wurzburg
Versbacher Landstrasse 7
WURZBURG 8700
Germany

C.A. Mims
Department of Microbiology
Guy's Hospital Medical School
London Bridge
LONDON, SE1 9RT
England

N. Nathanson
Institute for Experimental Pathology
University of Iceland
Keldur
REYKJAVIK
Iceland

H. Pabst
Institut für Virologie und Immunologie
der Universität Würzburg
Versbacher Landstrasse 7
WURZBURG 8700
Germany

P.A. Pálsson
Institute for Experimental Pathology
University of Iceland
Keldur
REYKJAVIK
Iceland

A. Pauloin
Institut National de la Santé et de la Recherche Médicale
u.-43, Hôpital Saint-Vincent-de-Paul
PARIS, 75014
France

G. Pétursson
Institute for Experimental Pathology
University of Iceland
Keldur
REYKJAVIK
Iceland

D. Sergiescu
Institut National de la Santé et de la Recherche Médicale
u.-43, Hôpital Saint-Vincent-de-Paul
PARIS, 74014
France

J.R. Stephenson
Institute of Virology and Immunobiology
University of Würzburg
Versbacherstrasse 7
8700 WURZBURG
Germany

L.Thiry
Head, Department of Virology
Institut Pasteur du Brabant
Rue du Remorqueur 28
1040 BRUSSELS
Belgium

H. Valdimarsson
Department of Immunology
St. Mary's Hospital Medical School
Praed Street
LONDON W2
England

R.A. Weiss
Imperial Cancer Research Fund Laboratories
P.O. Box 123
Lincoln's Inn Fields
LONDON WC2A 3PX
England

INTRODUCTION (C. A. MIMS)

There are two ways of looking at slow and persistent virus infections. The first is to consider their immense biological interest, whether or not they are of any practical importance. For instance, I could maintain that the most fascinating persistent virus of all is lactic dehydrogenase virus in mice. But this infection, in which only macrophages are involved and in which there are puzzling immunological phenomena, causes no pathological changes, no illness, and is of little or no importance for the mouse. The second approach to slow and persistent virus infections is a clinical one and reflects our eagerness to discover that viruses are behind this or that chronic disease of unknown aetiology. The neurologists, rheumatologists, and those who deal with cancer are interested from this point of view. The two approaches often overlap. SSPE, for instance, although it is a clinical problem, has a wider biological interest. We cannot understand oncornaviruses or visna virus without considering their relationship with the host genes and with the host species - in other words their general biology. Much of the scrapie work focuses unashamedly on scrapie as a fascinating problem in general biology but scrapie is also a practical problem in sheep.

It is a great pleasure to see that both the biological and the more clinical or practical aspects of these infections have been so neatly fitted into our programme. Its good also to see some immunology because immunology comes into everything, and you cannot understand any infectious process without looking at the immune response.

The rest of my short introductory talk consists of three points: First, there must be some more persistent viruses waiting to be discovered in man. I do not refer to C - type viruses, which for all I know have already been discovered in the form of nucleic acid sequences or virus - specific enzymes. But there are the papovaviruses, a fine set of persistent viruses, many of which are still what we used to call orphan viruses, looking for diseases. JC and BK viruses, excreted in the urine of transplant patients and pregnant women, infect most of us, and we need to learn more about them. But there appear to be other human viruses in this group, because non - BK non - JC viruses have also been isolated. Even the common wart virus has now been unequivocally divided into at least four distinct types by restriction enzyme analysis. Dr. Kalder at the San Antonio Primate Centre now has seven antigenically distinct simian foamy viruses. Surely there are some human foamy viruses. If so, then it is possible they have no effects on their host, in which case their biological interest is great but their practical importance zero. Are there representatives of visna virus in man? Were the reports of visna antibodies in human serum false alarms?

The second point is that we may discover that some of the old viruses do unexpected things. If human picornaviruses are capable of persisting or remaining latent like Theiler's virus in mice, it will raise many possibilities. Chronic infection with Theiler's virus sometimes causes an immunologically mediated demyelinating disease in mice. There have been attempts to find poliovirus RNA sequences in amyotrophic lateral sclerosis, but so far these have been unsuccessful. Even C - type viruses can be neurotropic, and one of them causes a chronic neurological disease in the mouse, probably by a direct effect of the virus rather than via the immune response. Yellow fever virus may seem an odd one to mention at such a meeting, but I have noticed how difficult it is to explain to immunologists how neutralizing antibodies to yellow fever remain at high levels for 50 - 70 years after the primary infection, when the virus was presumably eliminated from the body. Could it be that in some corner of the lymphoreticular system viral antigens persist, or there is a very slow turn over of productive infection? Hepatitis B virus certainly persists, but little is known about its ability to infect or remain latent in parts of the body other than the liver.

My last point is about viruses and the immune system. This is an area of research which seems full of opportunities. If a virus is to establish a persistent infection it must come to terms with immune responses, either by-passing them, avoiding them, or inducing ineffective responses. It can be no accident that nearly all persistent viruses, and also scrapie, go first to lymphoid tissues. To evade host defences, what more audacious but logical a strategy than to <u>invade</u> and in some way weaken these defences. There are various fascinating possibilities. We have suitable experimental techniques for dissecting out this interaction of viruses with lymphoreticular tissues, and by using the in vitro spleen cell system for instance it should be possible to discover a great deal that is relevant for persistence.

And now with great pleasure I will make way for those who have some hard data to present.

THE BIOLOGY OF SCRAPIE AGENT

H. KIMBERLIN

1.   TRANSMISSIBILITY OF SCRAPIE

Scrapie is a fatal disease of the CNS that occurs naturally
in sheep and goats (1). The clinical signs are variable
but affected animals have either incoordinated movements,
particularly in the hind limbs or show signs of intense
pruritis. Commonly, both types of abnormality occur. The
disease is diagnosed by clinical signs and the presence of
vacuolated nerve cells in histological sections of brain.
Interstitial spongy degeneration is often found in the same
areas as neuronal vacuolation and occasionally there may be
neuronal loss. Hypertrophy of astrocytes occurs as an
additional but non-specific lesion. Demyelination is
either very slight or absent and there are no inflammatory
changes to indicate the presence of an infectious agent
(Chapter 4 and reference 2).

However, there is no doubt that scrapie is caused by
a transmissible agent. The injection of brain homogenates
from affected sheep will transmit the disease to other
sheep after long incubation periods which sometimes last
for several years (1). The transmissible agent can be
filtered (3,4) and experimentally passaged in sheep to
extremely high dilutions of original inoculum (5) thus
demonstrating the existence of a replicating, virus-like
agent. Experimental forms of scrapie have been produced in
many species (Table 1), notably mice and hamsters. It is
important that several strains of mouse passaged agent have
been injected into sheep and produced scrapie (7). As
discussed later, scrapie is an infectious disease (section
3.3.1) and the causal agent shows the expected micro-
biological properties of strain variation (section 3.4) and

mutation (section 3.7).

Table I. Known susceptible hosts for experimental scrapie

Group	Species
Ruminant	Sheep, Goat
Carnivore	Mink
Old World Monkey	Cynomolgous
New World Monkey	Squirrel, Capuchin, Spider
Rodent	Mouse, Rat, Gerbil, Vole
	Hamster (Syrian and Chinese)

Adapted from reference 6

2. PHYSICOCHEMICAL PROPERTIES OF SCRAPIE AGENT

Despite intensive study, there is little firm information
on the nature of the scrapie agent (8). The only available
assay is by titration in animal hosts, which even in the
quickest model of scrapie (strain 263K in hamsters; 9)
takes 150-200 days. Infectivity titres accurately reflect
amounts of agent in inocula that are chemically similar but
the proportionality between titre and agent changes when
some chemical treatments are used, for example sodium
dodecyl sulphate (SDS) (10). This happens because highly
purified agent is not available and the non-scrapie
components in a tissue extract may become chemically
modified on treatment and, as a consequence, the efficiency
of infection is altered. Hence much of the published data
are difficult to interpret, particularly when infectivity
titres differ by only 1 to 2 $\log_{10}$ LD_{50} units.
   Most studies have been carried out with the 139A
strain of mouse passaged agent or with other strains from
the 'drowsy-goat' source. In retrospect this may have been
a mistake because there are some indications that biologi-

cally different strains of agent have different physico-
chemical properties. For example, the inactivation of the
22C strain of agent was about 3 $\log_{10}$ LD_{50} units greater
than that of the 22A agent when 10 percent saline homogenates of scrapie mouse brain were autoclaved at $110^{o}C$
for 30 minutes (11). Because of these findings it may be
premature to draw general conclusions about the nature of
the scrapie agent.

Another limitation of past work is that most of it
has been carried out with scrapie brains taken in the
clinical stage of the disease. Table 2 shows the results
of three preliminary experiments on the effects of SDS on
titre in scrapie brains taken at different times during
incubation. There is a clear pattern showing an apparent
increased inactivation of scrapie (strain 139A) at earlier
times than at later times. This pattern could be due to
structural differences between early and late synthesised
agent or, alternatively, to an alteration in brain tissue
as lesions develop in the second half of the incubation
period.

With these limitations in mind, the following is a
brief summary of the main findings on the nature of the
139A (Chandler) strain of agent. In general the agent is
highly stable when exposed to many physicochemical treatments, for example wet heat, alkylating agents, organic
solvents, concentrated salt solutions and many detergents
(8). This stability is probably related to the common
finding that infectivity is functionally associated with
cell membranes particularly in the microsome fraction. In
one study of the SMB cell line (12), derived from a
scrapie-affected brain and persistently infected with
agent, the highest infectivity titres were found in the
plasma membrane of the cell. Treatments which disaggre-
gate membrane structures, e.g. 80% 2-chloroethanol, 90%
phenol, 5% SDS (8,13), also appear to destroy most of the
scrapie infectivity, again suggesting a link between agent
and membranes. The agent has not been identified by the

Table 2. Effect of SDS on scrapie infectivity in brain homogenates prepared at different times in the incubation period

Days after i.c. infection with strain 139A	Infectivity titres ($-\log_{10}$ i.c. LD_{50} units/.03g)			
	Agent titre in brain	Loss of titre after treatment with SDS		
		Expt. 1	Expt. 2	Expt. 3
35	5.25		$\geqslant$ 2.50	
46	6.17	2.17		
49	6.21		2.14	
64	6.50		2.25	
68	7.27			3.25
76	7.00	1.17		
96	7.29			2.61
112	7.50		1.56	
117	7.33	1.00		
126	7.88		1.88	
138	7.88			2.41

Pooled mouse brains were homogenised in 0.32M sucrose at a concentration of 10% w/v and centrifuged at 1,000g for 10 min. to remove nuclei, myelin and unbroken cells. The supernatants were further centrifuged at 100,000g for 1 h. to sediment particulate material and most of the scrapie infectivity. The pellets were resuspended in saline at a concentration equivalent to 10% whole brain and aliquots were incubated with equal volumes of 1% SDS at pH 8.4. In experiments 1, 2 and 3, incubations were carried out at 20°C for 1 h., 37°C for 1 h. and 37°C for 2 h., respectively. Titrations of infectivity were performed on serial ten-fold dilutions injected intracerebrally (i.c.) into Compton white mice. Unpublished data of Kimberlin and Walker.

electron microscope and no one has purified scrapie
infectivity by more than 50-fold. Because of these results,
it has been suggested that scrapie agent may not exist as
a nucleoprotein structure of regular morphology.

The minimal operational size of the infectious scrapie
agent (assuming a spherical particle) has been estimated at
approximately 30 nm by membrane filtration (14), greater
than 5×10^7 molecular weight by gel filtration (14) and at
least 40 S by sedimentation properties (15). The target
size to ionising radiation is very small indeed, the
calculated molecular weight being about 10^5 (16,17,18).
This value is consistent with the very high doses of u.v.
irradiation needed to inactivate the agent (19) and it is
tempting to suggest the target to ionising radiation and
the chromophore for u.v. absorption is a scrapie nucleic
acid of about 10^5 molecular weight. However, if the
chromophore is nucleic acid then the unusual u.v. inactivation spectrum for scrapie infectivity (18,20) indicates
that it is associated with other macromolecules, possibly
protein. This suggestion is supported by the enhanced
inactivation of scrapie suspensions when exposed to ionising
radiation in the presence of oxygen (21). In summary, the
simplest interpretation of available data is that the agent
is made up of a small scrapie specific nucleic acid,
functionally associated with proteins (and perhaps other
macromolecules) to give an infectious complex which *in vivo*
is located in cell membranes.

Until recently there was no direct evidence for the
existence of a scrapie specific nucleic acid. However,
studies in which a hamster-passaged strain of agent was
exposed to DNAase have produced the first direct evidence
for a DNA component (22). This DNA has not yet been
identified but at least one possible candidate has been
described (23,24).

3.   THE NATURAL DISEASE

3.1.   *Contagious spread*

The results of three major studies, shown in Table 3, leave
no doubt that horizontal or contagious spread of scrapie
agent occurs naturally when previously unexposed sheep are
maintained with infected flocks.   Experiment II is
impressive because the exposed sheep came from a flock in
which no cases of scrapie had been seen in over 18,000
sheep during 13 years, and in experiment III, the numbers
of sheep exposed to scrapie and eventually developing the
disease were exceptionally large.

   Although the mode of transmission of agent is not
known for certain, several experimental routes of infection
have been identified which are likely to be routes of
natural infection.   These include oral dosing (28),
scarification (5) and via the conjunctiva (29).   Studies in
sheep and goats have failed to detect agent in secretions
but agent is present in many tissues (1).   Relatively large
amounts are present in foetal membranes (28) making this
tissue a probable source of infection.   Other, but untested
possibilities are cells sloughed off from mucous membranes
or damaged skin.

   There is some tentative evidence that sheep may
become infected by grazing infected pastures (30) and the
stability of scrapie agent to adverse conditions (section
3.2) makes this a plausible idea.   Persistence of agent in
the environment, on pastures or in buildings, is the likely
explanation for the failure to eradicate Rida (Icelandic
scrapie) from farms that were left free from sheep for at
least 1 year before restocking from Rida free areas; the
disease reappeared later but only on farms where it had
previously occurred (31).

3.2.   *Maternal transmission*

In general the probability of a lamb developing scrapie

Table 3. Contagious spread of scrapie

Experiment number	Date	Number of sheep or goats		Percent with scrapie	Reference
		Exposed	Developing scrapie		
I	1968	7 sheep	3	43	25
		17 goats	10	59	25
II	1974	75 sheep	21	28	26
III	1979	95 goats	50	53	27
		263 sheep	48	18	27

depends more on the eventual scrapie status of the ewe than of the ram (26,27,32). This bias indicates maternal transmission of the agent but the route by which this occurs is not firmly established. In part, maternal transmission can be explained by contagion during the months of intimate contact between ewe and lamb. However there is reasonable evidence that infection of the embryo occurs although this has not been shown directly. First, when ewes were experimentally infected with the SSBP/1 source of scrapie at the time of conception, some of the lambs developed scrapie at an exceptionally early age (4,33). This suggests pre-natal infection, particularly as there is independent evidence that lateral spread does not occur under these conditions (25). Secondly, there is the evidence mentioned above that foetal membranes from scrapie affected dams contain appreciable amounts of agent (28).

Several important questions concerning the spread of infection remain unanswered. For example, scrapie occurs most frequently in sheep between $2\frac{1}{2}$ and $4\frac{1}{2}$ years of age (1). A ewe could become infected before birth and could produce several lamb crops before she herself developed the disease. But it is not known how efficiently an infected ewe can transmit infection to her offspring at different stages during the incubation period.

3.3. *Host genetic factors*

Although scrapie is caused by an infectious agent there is clear evidence that host genetic factors are important to the development of disease. As discussed elsewhere (1,34) the nature of the genetic control of natural scrapie is ill defined and difficult to study. The most fundamental problem is that there is no means of assessing the degree of exposure of individual sheep to scrapie agent and hence, one cannot tell whether an animal fails to develop the disease because of its genes or because it never has a chance to become infected.

This problem is removed by infecting sheep experimentally. Lines of Cheviot and of Herdwick sheep have been bred for increased (positive line) or decreased (negative line) incidence of scrapie in response to a standard dose of the SSBP/1 source of agent (35,36). Line crossing experiments have shown that the response to experimental infection is mainly controlled by a single gene with the dominant allele conferring susceptibility (1,35).

The difference between positive and negative lines was seen in terms of the incidence of scrapie and in the length of incubation which was very long in the relatively few cases that occurred in negative line animals. By analogy with the genetic control of scrapie in mice (Sections 3.4.2 and 3.9), it seems likely that the genotype of sheep influences incubation period (which possibly exceeds natural life-span in some negative line sheep) by controlling the dynamics of agent replication.

Positive and negative line sheep that were selected against SSBP/1 have been injected with other strains of agent and in general the responses were predictable (7).

However, at least one strain of agent is known (CH1641) which produces scrapie in both lines of Cheviot sheep with similar incidences but with shorter incubation periods in the negative line (7). This variation in response with agent strain is reminiscent of the diversity of inter-actions between strains of mouse passaged agent and the alleles of *Sinc* gene (Section 3.4.2) and it complicates the possible use of genetical methods of controlling the natural disease (34). The genetic control of natural scrapie may be even more complex than is indicated by these experiments because additional genetic factors may control stages of infection that are by-passed when sheep are infected experimentally. Hence, the genetic control of scrapie could vary with the predominant route of natural infection as well as with the strain of agent (1).

3.4. *Control*

Without an efficient diagnostic test to identify infected animals it would seem that eradication of endemic scrapie is impossible at the moment. Selective breeding may one day be a useful method of control in certain circumstances but meanwhile, some kind of slaughter policy is the only practicable way of coping with natural scrapie (1,34). The slaughter of both affected flocks and source flocks is an effective method of control. Less drastic measures depend on limiting the maternal and lateral spread of infection by selective culling of the female line of scrapie cases and by such measures as the removal of after-births at lambing.

3.5. *Related diseases*

Scrapie is the best understood of a group of four slow transmissible encephalopathies, the other three being kuru,

Creutzfeldt-Jakob disease (CJD) and transmissible mink encephalopathy (TME). The main link between these diseases is the broadly similar pattern of histological lesions found in brain (6,37). In addition, some combinations of scrapie strain and mouse genotype are associated with cerebral amyloid plaques which are a conspicuous feature of Alzheimer's disease and other dementias (38,39). Hence scrapie is an important model of some human CNS diseases. This does not necessarily imply that scrapie agent is the cause of human disease, but the possibility should be considered.

TME is a rare disease and epidemiological studies have pointed towards a source of agent outside the mink population (40,41). There is evidence that the source of TME agent is scrapie infected carcasses and that the natural route of infection is via bite wounds inflicted by littermates, particularly at feeding times (42). It has been postulated that the occurrence of CJD may also be related to the ingestion of scrapie contaminated food (43). However the epidemiological evidence conflicts with this view because the world-wide distribution of CJD does not correlate with the distribution of sheep or with the occurrence of scrapie (34). It may be nearer the truth to think of the CJ agent as a relatively common infection in the human population which only very rarely produces disease.

4. IDENTIFICATION OF AGENT STRAINS

4.1. *Lesion profiles*

The absence of serological or cell culture methods for identifying different strains of scrapie agent has led to the development of alternative techniques. One of these is based on a quantitative assessment of the severity of

vacuolar lesions in 9 grey matter areas and 3 white matter
areas of mouse brain (2). The resulting lesion profile
is influenced by many variables but if these are carefully
controlled it can be used to identify different strains
of agent (37). The technique is very sensitive because
many agents produce characteristic but different profiles
in mice of different *Sinc* genotypes (see section 3.4.2).
In addition, there are qualitative differences between
agent strains. For example, strains 79A and 22C are rarely
if ever associated with cerebral amyloidosis whereas 87A
and 87V produce many amyloid plaques and show some
asymmetric vacuolation in certain areas of the brain (37,
44).

4.2. *Biological properties*

The incubation period of scrapie in mice is under the
control of a gene, *Sinc* which has two alleles, s7 and p7.
One group of agent strains behaves like ME7 and has a
relatively short incubation period in mice with the *Sinc*
genotype s7s7 and a much longer incubation period in mice
homozygous for *Sinc*p^7. Another group, exemplified by 22A,
shows the reverse properties (46). However many members of
both the ME7 and 22A groups can be distinguished when the
incubation periods in the heterozygotes are compared with
those found in the parental, homozygous genotypes. For
example, within the ME7 group of agents, ME7 itself shows
no dominance, 22C shows partial dominance of the s7 allele,
79A, partial dominance of p7 and 139A, overdominance of p7
(47,48). This remarkable diversity of interactions between
strains of agent and the alles of *Sinc* gene is not only of
great theoretical importance (section 3.9) but also
provides a way of identifying different strains which is
independent of the lesion profile system described above.
The combined use of these techniques, i.e. the comparison
of incubation periods and lesion profiles in mice of the

three *Sinc* genotypes has led to the identification of about
15 different strains of scrapie agent (7). The fastest
combinations of agent strain and mouse genotype produce
clinical scrapie after about 100-150 days but other combinations have incubation periods that exceed the natural
lifespan of laboratory mice (49).

5.   ISOLATION OF AGENT STRAINS IN MICE

Isolations are usually made by injecting homogenates of
sheep brain intracerebrally into mouse strains of the *Sinc*
genotypes s7s7 and p7p7. Single strains of agent can be
obtained by repeated intracerebral passage in mice or
better still, by passage at limiting infectious dilutions
of scrapie brain, i.e. cloning. It is known that several
strains of agent may be involved in a single outbreak of
scrapie. More important, two or more strains are frequently
isolated from individual sheep (1).

However, primary isolations of scrapie agent are
often difficult to make and there are some strains, for
example CH1641, which cannot be transmitted to mice. The
common finding is that incubation periods at first passage
in mice are greatly extended and more variable than those
found at second passage. This phenomenon is known as the
species barrier effect and its existence suggests that some
selective pressures may operate at primary passage which
influence the strains of agent isolated. Some of the
processes that take place on crossing the species barrier
are discussed more fully later on (section 3.6) and strain
selection is one of them.

In recent studies, scrapie agent has been isolated
from several natural cases taken from the same outbreak.
Table 4 shows that one group of sheep brains gave relatively
short incubation periods at first passage, in the range of
470 to 479 days. Those from a second group produced
scrapie in mice after 536 to 588 days, and in a single

Table 4. Mean incubation periods at first passage in mice
of scrapie agent from natural cases

Individual sheep	Incubation periods in mice (days ± SEM)
C 58*	470 ± 13
M575	471 ± 9
F 47*	479 ± 11
G137	536 ± 26
H 77	561 ± 7
F 70*	585 ± 14
F 85*	588 ± 19
A 8*	671 ± 21
Pool	472 ± 11

The sheep came from the same flock of Herdwicks kept at
Compton and were clinically and histologically positive
for scrapie. Brain homogenates (5 or 10% w/v) were
prepared in saline and 0.03 ml of each were injected intracerebrally into groups of 16 Compton white mice ($Sinc^{s7}$).
*Denotes individual sheep brains used to make up the pool.
Unpublished data of Kimberlin and Walker.

case, the mean incubation period was 671 days (Table 4).
These preliminary results suggest that 2 or 3 different
strains of agent were involved in the outbreak. However
the injection of a representative pool of these sheep
brains gave an incubation period of 472 days, indicating
the selection of a quicker agent (Table 4).
    The second passage was carried out using 5-8
individual mouse brains taken from the first passage of
each sheep source of scrapie agent. The second passage
from individual sheep brains was extremely variable with
some incubation periods in the range of 139-163 days but

Table 5. Variation in incubation periods at second
passage in mice of scrapie agent from natural cases

Individual sheep	No. of 1st passage mouse brains tested	Mean incubation periods at second passage in mice (days)				
C58	5	147	150	189	358	395
F47	5	145	146	148	181	193
F70	6	163 391	227	239	243	249
F85	6	139 286	142	179	203	253
A 8	6	139	190	429	477	
Pool of above	8	150 175	151 176	154 248	155	165

Brain homogenates (1% w/v) were prepared from individual
mice with clinical scrapie at the first passage (Table 4)
and 0.03 ml was injected intracerebrally into groups of
16 Compton white mice to initiate the second passage. The
standard error of the mean incubation period was always
less than 5 percent and usually less than 2 percent.
Incubation periods of 176 days or less are underlined (see
text). Unpublished data of Kimberlin and Walker.

others were two or three times longer (Table 5). In
contrast, the second passage from the pool of sheep brains
was far less variable with all but one of the incubation
periods falling in the range of 150-176 days. These
findings support the conclusion drawn from the data in

Table 4, namely that different strains of scrapie agent
were involved in the outbreak and that isolation of agent
from pooled sheep brains tended to select a quicker strain
(in $Sinc^{s7}$ mice).

The results in Table 5 also indicate that both
'short' and 'long' incubation strains of agent seem to be
present in some sheep brains (e.g. C58 and A8) and that
they can be isolated by passage in a single strain of
mouse (Compton white; $Sinc^{s7}$). If this is the case then it
is not immediately obvious why the long incubation strains
of agent were isolated at all since the shorter ones should
have had the selective advantage. Possible explanations
could invoke differences between strains in their relative
concentrations in brain and in the efficiency with which
they infected mice at first and second passage. Alternatively there may have been blocking of one strain by
another, or perhaps mutations occurred on passage to give
variants with different incubation properties. Competitive
interactions are known to occur when different strains of
mouse passaged agent are injected into the same mice at
different times (section 3.8) but there is no direct
evidence at the moment that agents in the same inoculum can
compete. However, there is good evidence that scrapie
strains can mutate (section 3.7).

6.   THE SPECIES BARRIER

The most detailed studies of the species barrier effect
have been carried out with hamsters and mice and three
underlying processes have been identified (50).

Although there is no specific immune response to
scrapie infection (51), the intimate association between
scrapie agent and tissue components makes it likely that
there is an immune response to the antigenically foreign
inoculum which could lead to a reduced efficiency of
infection. Two studies have been made showing that this is

probably true. First, pretreatment of mice with normal
hamster brain decreased the proportion of scrapie cases
and increased incubation times following infection with
hamster passaged scrapie agent (52). In the second study,
mouse passaged ME7 agent was transmitted to sheep and then
reisolated in mice. On reisolation, the incubation period
of ME7 was extended at first passage in mice relative to
the second but the increase was exactly in accord with the
lower estimated titre in mice of the sheep passaged agent
(53,54).

However a reduced efficiency of infection is an
inadequate explanation of the species barrier effect when
the incubation period at first pass falls outside the
normal dose-response range of the agent in the new host.
It should also be emphasised that sometimes the species
barrier effect is minimal as when hamster passaged 431K is
transmitted to mice ($Sinc^{s7}$) (55).

The second process associated with interspecific
passage of scrapie has already been mentioned, namely
strain selection. In one study it was found that the early
passages of a source of scrapie agent in hamsters were
associated with the loss of strain 431K which had a longer
incubation period than strain 263K (55).

In the third process a prolonged 'zero-phase' occurs
at first passage which is either absent or much reduced on
second passage in the new host (56). The term 'zero-phase'
is used to describe the period of time before agent
replication can be detected, and in mouse passaged scrapie
its duration is controlled by $Sinc$ gene (section 3.9). In
one study of the transmission of hamster scrapie to mice it
was found that the dynamics of agent replication in brain
at first and second passage were almost identical so that
the difference in incubation periods was entirely accounted
for by a prolonged zero-phase at first passage (56). This
is summarised by the following equation:-

Incubation period   minus   Zero-phase   =   Incubation period
      1st pass   1st pass   2nd pass

    (325 days)   minus   (175 days)   =   (149 days)

The reason for such a prolonged zero-phase at first
passage is not known. In part it may be due to a decreased
efficiency of infection but other factors must also be
involved, for example, re-routing of hamster associated
agent along different and less efficient pathways.

7.   BIOLOGICAL STABILITY OF AGENTS

There is now strong evidence that some strains of scrapie
agent can undergo mutation on passage in mice (53) and
Table 6 gives a list of the stability classes found so far.
With class II agents, there is a gradual change in properties which is indicative of accumulated point mutations
over several passages. Agent strains in class III are
different in that the change in properties is unpredictable
and discontinuous. Class III agents include 87A and other
strains that are associated with a high incidence of
amyloid plaques in brain and with asymmetrical vacuolation.
When these agents change, the new or modified strains
appear to be the same in each case. This regularity
suggests either that mutations or deletions occur at the
same site or that the host selects strongly in favour of
one strain if many are produced. Probably both factors
determine the uniformity of the resulting strain.

8.   COMPETITION BETWEEN AGENTS

There is considerable evidence for competitive interactions
between some strains of agent (47). The basic experiment
is to inject mice with an agent strain (blocking agent)

Table 6. Biological stability of different classes of
scrapie agent

Class	Example	Characteristics
I	ME7	Completely stable on passage in either $Sinc^{s7}$ or $Sinc^{p7}$ mice.
II	22A	Stable on passage in the $Sinc$ genotype in which they were isolated but properties change on passage in mice of the other homozygous genotype.
III	87A	Unstable even on serial passage in the mouse genotype in which they were isolated.

Information taken from reference 53.

that is operationally slow in the genotype used, and after
an interval, to inject an agent that is quicker (killing
agent). Competition or blocking is seen as an increased
incubation of the second agent which can be identified by
its lesion profile. The original experiments (57,58) were
carried out with agents from the ME7 and 22A groups, and by
changing the genotype of mouse the 'blocking' and 'killing'
roles of the agents could be reversed. However, blocking
has now been demonstrated between agents within the ME7
group (48). Blocking can be achieved using either the
intracerebral or the intraperitoneal route and it is
independent of the mouse genotype in which the agents were
passaged (47,48).

The simplest and most convincing explanation of
blocking is that the first agent occupies a proportion of a
finite and relatively small number of agent replication

sites so that fewer are available when the second agent is injected. As a consequence the effective dose of the second agent is reduced and incubation is prolonged (48). In some cases blocking may be total (58). It is important that the extent of blocking can be precisely controlled by varying the respective doses of the two agents. For example, blocking is greater if the dose of the first agent is increased and that of the second agent reduced. Blocking is also greater if the interval between injections is increased because this allows more time for the slower, blocking agent to replicate and occupy sites that would otherwise be used by the killing agent (47,48).

There is now good evidence that blocking only occurs with an agent that under the conditions of the experiment can replicate. For example, a dose of agent that is too low to produce infection will not block (48). No blocking has been demonstrated with the related transmissible mink encephalopathy agent which is not transmissible to any of the 14 strains of mice tested (48). Neither does blocking occur with strain 22A that has been inactivated by chemical or physical treatment (59).

9.   REPLICATION SITE HYPOTHESIS FOR SCRAPIE

The pathogenesis of scrapie is discussed in detail in the next Chapter but it is necessary at this stage to emphasise some of the main features.

With the short incubation models of scrapie, agent replication occupies most of the incubation period (48,50, 51). Following peripheral routes of infection, agent replicates rapidly in spleen to reach plateau concentrations of infectivity within a few weeks. Agent also replicates in other lymphoid organs but spleen is known to play an important role because its removal lengthens incubation period (60-63). At a later stage, agent enters the CNS and a prolonged period of replication follows to give

concentrations of agent in the CNS that are much higher
than those found elsewhere. It is significant that histological lesions of scrapie only develop in the CNS (2) and
that the sharp onset of clinical disease appears to be
triggered by a certain critical concentration of agent in
brain or in, as yet unidentified, areas of brain (64). It
is also significant that agent replication occurs without
the apparent stimulation of a specific immune response to
infection and without the involvement of interferon (see
34,51). However the absence of a continuing host response
is consistent with the remarkable predictability of
incubation period which is a general feature of the
pathogenesis of scrapie.

The same sequence of events occurs in the long
incubation models of scrapie except that there is a prolonged zero-phase before agent replication begins and the
duration of the zero-phase in mice is precisely controlled
by *Sinc* gene (48,49,65,66). It is clear, therefore that
agent replication is the central process in the pathogenesis of mouse scrapie and that the slowness and predictability of disease development is due to an overiding
genetic control exerted by the host. Since genetic
factors influence the development of scrapie in sheep
(section 3.3.3) it is reasonable to suggest that genes
equivalent to *Sinc* exert a similar control on the natural
disease.

The interaction between strains of scrapie agent and
the alleles of *Sinc* gene described earlier (section 3.4.2)
is of great theoretical importance as well as being of
practical value in strain-typing. In particular, the
occurrence of overdominance indicates that the two alleles
of *Sinc* do not act independently and Dickinson has
suggested that each allele contributes a monomer to a
multimeric structure which is involved in agent replication
(46-48). These suggestions have been developed into the
Scrapie Replication Site Hypothesis which states that the
slow development of clinical scrapie in mice is due to a

restriction in the number of agent replication sites which
are multimeric structures specified by the alleles of *Sinc*
gene.

There is a considerable amount of experimental data
to support this hypothesis, and so far, none to refute it.
The most important evidence comes from the many studies
described earlier (section 3.8) showing competition
between strains of agent. The concept of a limited number
of agent replication sites is further supported by the
occurrence of a plateau concentration of agent in spleen
(45,49,61,67). The important role of the spleen in
pathogenesis probably arises because this organ contains a
high proportion of the peripheral replication sites in
mice. Splenectomy removes these sites which do not appear
to be replaced (63) so that agent has to replicate in a
smaller number of sites and incubation period is prolonged.
The hypothesis predicts that when the dose of injected
agent is less than the number of replication sites in
spleen, splenectomy will have no effect and the estimated
infectivity titres in splenectomised and intact mice will
be the same. Both these predictions have been fulfilled
(48).

It is obvious from this discussion that our detailed
understanding of scrapie pathogenesis is still very incomplete. However, studies of the biology of scrapie
agent have produced a well tested hypothesis which can
explain why scrapie is a slow disease and they have
identified some of the key questions to be answered by
future research. One of these is, what is the normal
function of *Sinc* gene?

REFERENCES

1. Dickinson, AG: Scrapie in sheep and goats. In: Slow virus diseases of animals and man, Kimberlin RH (ed), Amsterdam, North Holland, 1976, p 209-241.
2. Fraser, H: The pathology of natural and experimental scrapie. In: Slow virus diseases of animals and man, Kimberlin, RH (ed), Amsterdam, North Holland, 1976, p 267-305.
3. Wilson, DR, RD Anderson, W Smith: Studies in scrapie. J Comp Pathol 60: 267-282, 1950.
4. Gordon, WS: Review of work on scrapie at Compton, England, 1952-1964. In: Report of scrapie seminar, ARS 91-53, Washington, DC, US Department of Agriculture, 1966, p 8-40.
5. Stamp, JT, JG Brotherston, I Zlotnik, JMK Mackay, W Smith: Further studies on scrapie. J Comp Pathol 69: 268-280, 1959.
6. Marsh, RF: The subacute spongiform encephalopathies. In: Slow virus diseases of animals and man, Kimberlin, RH (ed), Amsterdam, North Holland, 1976, p 359-380.
7. Dickinson, AG, H Fraser: An assessment of the genetics of scrapie in sheep and mice. In: Slow transmissible diseases of the nervous system, Hadlow, WJ, SB Prusiner (eds), New York, Academic Press, in press.
8. Millson, GC, GD Hunter, RH Kimberlin: The physico-chemical nature of the scrapie agent. In: Slow virus diseases of animals and man, Kimberlin, RH (ed), Amsterdam, North Holland, 1976, p 243-266.
9. Kimberlin, RH, CA Walker: Characteristics of a short incubation model of scrapie in the golden hamster. J Gen Virol 34: 295-304, 1977.
10. Kimberlin, RH: Biochemical approaches to scrapie research. Trends in Biochemical Sciences 2: 220-223, 1977.
11. Dickinson, AG, DM Taylor: Resistance of scrapie agent to decontamination. N Engl J Med 299: 1413-1414, 1978.
12. Clarke, MC, GC Millson: The membrane location of scrapie infectivity. J Gen Virol 31: 441-445, 1976.
13. Millson, GC, EJ Manning: The effect of selected detergents on scrapie infectivity. In: Slow transmissible diseases of the nervous system, Hadlow, WJ, SB Prusiner (eds), New York, Academic Press, in press.

14. Kimberlin, RH, GC Millson, GD Hunter: An experimental examination of the scrapie agent in cell membrane mixtures. III. Studies of the operational size. J Comp Pathol 81: 383-391, 1971.

15. Prusiner, SB, WJ Hadlow, CM Eklund, RE Race, SP Cochran: Sedimentation characteristics of the scrapie agent from murine spleen and brain. Biochemistry 17: 4987-4992, 1978.

16. Alper, T, DA Haig: Protection by anoxia of the scrapie agent and some DNA and RNA viruses irradiated as dry preparations. J Gen Virol 3: 157-166, 1968.

17. Field, EJ, F Farmer, EA Caspary, G Joyce: Susceptibility of scrapie agent to ionizing radiation. Nature (London) 222: 90-91, 1969.

18. Latarjet, R: Inactivation of the agents of scrapie, Creutzfeldt-Jakob disease and kuru by radiations. In: Slow transmissible diseases of the nervous system, Hadlow, WJ, SB Prusiner (eds), New York, Academic Press, in press.

19. Alper, T, DA Haig, MC Clarke: The exceptionally small size of the scrapie agent. Biochem Biophys Res Commun 22: 278-284, 1966.

20. Latarjet, R, B Muel, DA Haig, MC Clarke, T Alper: Inactivation of the scrapie agent by near monochromatic ultraviolet light. Nature (London) 227: 1341-1343, 1970.

21. Alper, T, DA Haig, MC Clarke: The scrapie agent: evidence against its dependence for replication on intrinsic nucleic acid. J Gen Virol 41: 503-516, 1978.

22. Marsh, RF, TG Malone, JS Semancik, W Lancaster, RP Hanson: Evidence for an essential DNA component in the scrapie agent. Nature (London) 275: 146-147, 1978.

23. Corp, CR, RA Somerville: Nucleic acids associated with detergent-treated synaptosomal plasma membranes from normal and scrapie-infected mouse brain. Biochemical Soc Trans 4: 1110-1112, 1976.

24. Hunter, GD: The enigma of the scrapie agent: biochemical approaches and the involvement of membranes and nucleic acids. In: Slow transmissible diseases of the nervous system, Hadlow WJ, SB Prusiner (eds), New York, Academic Press, in press.

25. Brotherston, JG, CC Renwick, JT Stamp, I Zlotnik, IH Pattison: Spread of scrapie by contact to goats and sheep. J Comp Pathol 78: 9-17, 1968.

26. Dickinson, AG, JT Stamp, CC Renwick: Maternal and lateral transmission of scrapie in sheep. J Comp Pathol 84: 19-25, 1974.

27. Hourrigan, JL, AL Klingsporn, WW Clarke, M de Camp: Epidemiology of scrapie in the United States. In: Slow transmissible diseases of the nervous system, Hadlow, WJ, SB Prusiner (eds), New York, Academic Press, in press.

28. Pattison, IH, MN Hoare, JN Jebbett, WA Watson:
Further observations on the production of scrapie in
sheep by oral dosing with foetal membranes from
scrapie-affected sheep. Br Vet J 130: lxv-lxvii,
1974.

29. Haralambiev, H, I Ivanova, A Vesselinova, K Mermerski:
An attempt to induce scrapie in local sheep in
Bulgaria. Zentralbl Veterinaermed B 20: 701-709,
1973.

30. Greig, JR: Scrapie in sheep. J Comp Pathol 60:
263-266, 1950.

31. Pálsson, PA: Rida: its epidemiology in Iceland. In:
Slow transmissible diseases of the nervous system,
Hadlow, WJ, SB Prusiner (eds), New York, Academic
Press, in press.

32. Dickinson, AG, GB Young, JT Stamp, CC Renwick: An
analysis of natural scrapie in Suffolk sheep.
Heredity (Lond) 20: 485-503, 1965.

33. Dickinson, AG, GB Young, CC Renwick: Scrapie:
experiments involving maternal transmission in sheep.
In: Report of scrapie seminar, ARS 91-53, Washington
DC, US Department of Agriculture, 1966, p 244-248.

34. Kimberlin, RH; Scrapie as a model slow virus disease:
problems, progress and diagnosis. In: Comparative
diagnosis of viral diseases. Vol III. Animal and
related viruses, Kurstak, E, C Kurstak (eds),
New York, Academic Press, in press.

35. Dickinson, AG, JT Stamp, CC Renwick, JC Rennie: Some
factors controlling the incidence of scrapie in
Cheviot sheep injected with a Cheviot-passaged
scrapie agent. J Comp Pathol 78: 313-321, 1968.

36. Nussbaum, RE, WM Henderson, IH Pattison, NV Elcock,
DC Davies: The establishment of sheep flocks of
predictable susceptibility to experimental scrapie.
Res Vet Sci 18: 49-58, 1975.

37. Fraser, H: Neuropathology of scrapie: the precision
of the lesions and their diversity. In: Slow
transmissible diseases of the nervous system,
Hadlow, WJ, SB Prusiner (eds), New York, Academic
Press, in press.

38. Bruce, ME, AG Dickinson, H Fraser: Cerebral amyloidosis
in scrapie in the mouse: effect of agent strain and
mouse genotype. Neuropathol Appl Neurobiol 2:
471-478, 1976.

39. Wisniewski, HM, ME Bruce, H Fraser: Infectious
etiology of neuritic (senile) plaques in mice.
Science 190: 1108-1110, 1975.

40. Hartsough, GR, D Burger: Encephalopathy of mink.
I. Epizootiological and clinical observations. J
Infect Dis 115: 387-392, 1965.

41. Burger D, GR Hartsough: Encephalopathy of mink. II.
Experimental and natural transmission. J Infect Dis
115: 393-399, 1965.

42. Marsh, RF, RP Hanson: On the origin of transmissible
    mink encephalopathy. In: Slow transmissible diseases
    of the nervous system, Hadlow, WJ, SB Prusiner (eds),
    New York, Academic Press, in press.
43. Gajdusek, DC: Unconventional viruses and the origin
    and disappearance of kuru. Science 197: 943-960,
    1977.
44. Fraser, H: Scrapie: a transmissible degenerative CNS
    disease. In: Progress in neurological research,
    Behan, PO, F Clifford Rose (eds), London, Pitman,
    in press.
45. Dickinson, AG, VMH Meikle, H Fraser: Identification of
    a gene which controls the incubation period of some
    strains of scrapie agent in mice. J Comp Pathol 78:
    293-299, 1968.
46. Dickinson, AG, VMH Meikle: Host-genotype and agent
    effects in scrapie incubation: change in allelic
    interaction with different strains of agent. Mol Gen
    Genet 112: 73-79, 1971.
47. Dickinson, AG, H Fraser: Scrapie: pathogenesis in
    inbred mice: an assessment of host control and
    response involving many strains of agent. In: Slow
    virus infections of the central nervous system,
    ter Meulen, V, M Katz (eds), New York, Springer-Verlag,
    1977, p 3-14.
48. Dickinson, AG, GW Outram: The scrapie replication-site
    hypothesis and its implications for pathogenesis. In:
    Slow transmissible diseases of the nervous system.
    Hadlow, WJ, SB Prusiner (eds), New York, Academic
    Press, in press.
49. Dickinson, AG, H Fraser, GW Outram: Scrapie incubation
    time can exceed natural lifespan. Nature (London)
    256: 732-733, 1975.
50. Kimberlin, RH: Early events in the pathogenesis of
    scrapie in mice: biological and biochemical studies.
    In: Slow transmissible diseases of the nervous
    system. Hadlow, WJ, SB Prusiner (eds), New York,
    Academic Press, in press.
51. Outram, GW: The pathogenesis of scrapie in mice. In:
    Slow virus diseases of animals and man, Kimberlin RH
    (ed), Amsterdam, North Holland, 1976, p 325-357.
52. Kimberlin, RH, CA Walker, GC Millson: Interspecies
    transmission of scrapie-like diseases. Lancet ii:
    1309-1310, 1975.
53. Bruce, ME, AG Dickinson: Biological stability of
    different classes of scrapie agent. In: Slow
    transmissible diseases of the nervous system,
    Hadlow, WJ, SB Prusiner (eds), New York, Academic
    Press, in press.
54. Dickinson, AG: personal communication.
55. Kimberlin, RH, CA Walker: Evidence that the trans-
    mission of one source of scrapie agent to hamsters
    involves separation of agent strains from a mixture.
    J Gen Virol 39: 487-496, 1978.

56. Kimberlin, RH, CA Walker: Pathogenesis of scrapie:
    agent multiplication in brain at the first and
    second passage of hamster scrapie in mice. J Gen
    Virol 42: 107-117, 1979.
57. Dickinson, AG, H Fraser, VMH Meikle, GW Outram:
    Competition between different scrapie agents in mice.
    Nature (New Biol) 237: 244-245, 1972.
58. Dickinson, AG, H Fraser, I McConnell, GW Outram,
    DI Sales, DM Taylor: Extraneural competition between
    different scrapie agents leading to loss of
    infectivity. Nature (London) 253: 556, 1975.
59. Kimberlin, RH, AG Dickinson: unpublished.
60. Fraser, H, AG Dickinson: Pathogenesis of scrapie in
    the mouse: the role of the spleen. Nature (London)
    226: 462-463, 1970.
61. Clarke, MC, DA Haig: Multiplication of scrapie agent in
    mouse spleen. Res Vet Sci 12: 195-197, 1971.
62. Dickinson, AG, H Fraser: Scrapie: effect of Dh gene
    on incubation period of extraneurally injected agent.
    Heredity 29: 91-93, 1972.
63. Fraser, H, AG Dickinson: Studies of the lymphoreticular
    system in the pathogenesis of scrapie: the role of
    spleen and thymus. J Comp Pathol 88: 563-573, 1978.
64. Kimberlin, RH, CA Walker: Pathogenesis of mouse
    scrapie: effects of route of inoculation on
    infectivity titres and dose-response curves. J Comp
    Pathol 88: 39-47, 1978.
65. Dickinson, AG, H Fraser: Genetical control of the
    concentration of ME7 scrapie agent in mouse spleen.
    J Comp Pathol 79: 363-366, 1969.
66. Dickinson, AG, VMH Meikle, H Fraser: Genetical control
    of the concentration of ME7 scrapie agent in the
    brain of mice. J Comp Pathol 79: 15-22, 1969.
67. Hunter, GD, RH Kimberlin, GC Millson: Absence of
    eclipse phase in scrapie mice. Nature (New Biol)
    235: 31-32, 1972.

THE PATHOGENESIS AND PATHOLOGY OF SCRAPIE

H. FRASER

1.   INTRODUCTION

Although scrapie represents a major source of loss to the
sheep industry, it has become increasingly apparent that
its study can provide insight into an important group of
enigmatic human neurological diseases. In addition
however, it has a peculiar fascination for the biological
scientist as it represents a uniquely perplexing subject
as far as most current views on infectious processes are
concerned.

        The disease results from a progressive
degeneration of the neuroparenchyma, which it has not been
found possible to reverse.   At present there is almost
complete ignorance concerning the molecular basis of this
deterioration, despite the large amount of relevant data
which is now available.   Scrapie is a naturally occurring
disease in sheep and goats, but it is studies in
laboratory rodents which are providing the means of
unravelling its basic biology.   The development of a
procedure using inbred mice, for the isolation and
discrimination of the many different strains of scrapie
agent represents the single most significant advance in
recent years.   These studies have also identified a wide
range in the manifestation of the disease, whose basis
has become clearly established as a result of the
experimental work with the inbred mice.   This has led to
a recognition that several human diseases have a variety
of features in common with one or other of the various
"types" of scrapie, and it cannot be anticipated where
such similarities will end, because there is no reason to
think that the full range of scrapie has yet been found.
It is easy to recognise the analogy with Creutzfeldt-Jakob
(C-J) disease and kuru, both of which have been assigned,

with scrapie and transmissible mink encephalopathy (TME) to a group of infectious degenerative encephalopathies. In addition, in the case of Alzheimer's disease, scrapie clearly provides a highly appropriate model for a variety of reasons (1) even if Alzheimer's disease does not turn out to have an infectious aetiology.

Although scrapie is unquestionably an infectious disease, both experimentally and in the field situation, there remain a number of issues where knowledge of it departs from expectations for conventional infectious processes. Firstly one of the most serious limitations in its identification and study is that it can only be recognised by clinical and pathological criteria, and all experimental work is completely dependent upon whole animal systems, usually in mice in which clinical disease occurs only after long incubation periods sometimes approaching or even exceeding median life span. In hamsters a "rapid" incubation period model has been developed, which has advantages for certain types of investigation (2). There is still no in vitro diagnostic recognition test for scrapie in infected or affected animals. Secondly it has not yet been possible to achieve biochemical purity of the agents of scrapie, and all transmissions and pathogenesis studies are therefore dependent upon more or less crude homogenates, usually of brain or spleen, but containing a wide range of tissue components and degeneration products. Thirdly it is not always possible even to transmit the natural disease to mice from all cases, the failure rate being around 20% (3), although this may to some extent reflect the limits imposed by the life span of the species concerned (4). An analogous situation has been found experimentally in which a scrapie agent passaged in hamsters, originally from a mouse isolate, has lost its infectivity for mice (5). Fourthly it has not been possible to recognise any scrapie agents ultrastructurally, and fifthly those agents that have been studied extensively do not express

antigenicity, although many of the more recently
identified agents, such as those responsible for the
induction of cerebral amyloid, have not been studied in
this respect.   Finally it should not be forgotten that
it has not yet been possible to show that Koch's third
postulate has been formally met with any of the agents
isolated, by reproducing all the features of the natural
disease in the species of origin.

A major difference in the behaviour of scrapie,
with wide ranging epidemiological significance, is that,
whereas it is both contagious in sheep and transmitted
vertically from mother to offspring, neither of these
phenomena have been shown to occur experimentally in mice.

2.   PATHOGENESIS

Although scrapie is a neurological disease, resulting from
changes induced in the central nervous system (CNS) many
months after initial infection, this is preceded by
replication of agent in several peripheral organs, and the
only way of altering the progress of the infectious
process is by interfering with it either during the
prelude of the infection, when the early peripheral events
are commencing, or during the peripheral replication but
prior to CNS involvement.   Once the CNS infection has
become established it has not been possible to alter its
progress in any way.   If the initial infection is directly
into the CNS, as, for example, with the intracerebral
route of injection, replication proceeds there
independently of replication elsewhere, and under these
circumstances it has not been possible to impede or alter
the progress of the disease at any stage.

The rate of agent replication in mice and the
timing of events such as the onset of the clinical
disease are all controlled by a single host gene
called Sinc (scrapie incubation period).   The action of
Sinc distinguishes agents in a way which has now become

an important and established procedure in their
classification and identification. Two alleles, $\underline{Sinc}^{s7}$
and $\underline{Sinc}^{p7}$, have been identified, and the dozen or so
agents which are currently known can be divided broadly
into an "ME7 group", which have shorter incubation periods
in $\underline{Sinc}^{s7}$ homozygotes, and a "22A group" with shorter
incubation periods in the $\underline{Sinc}^{p7}$ homozygotes (6). A
similar gene with two alleles $\underline{Sip}$ and $\underline{sip}$ has been
identified in Cheviot sheep, based on their susceptibility
to a particular experimental challenge agent (3, 7). On
the basis of the differences in $\underline{Sinc}$ action, the
$\underline{replication\ site\ hypothesis}$ has been proposed, which
suggests that the gene products of $\underline{Sinc}$ associate as
multimers and these are restricted in number (the
multimers will include heteromers in the heterozygote).
This hypothesis successfully explains many of the peculiar
features of the pathogenesis of scrapie (6,8).

The absolute timing of the events controlled by
$\underline{Sinc}$ is directly influenced by the number of infectious
units of infectivity introduced at the initial infection,
and so the five major variables which interact to produce
the final phenotypic expression of incubation period are:
1. host genotype; 2. strain of agent; 2. route of
inoculation; 4. dose of agent; 5. immunological status
(including developmental maturity). It is possible to
choose a combination of these variables such that a "very
short" incubation period disease can be produced, "very
short" being something in the region of 20 weeks in mice
(the details will depend upon the precise criteria for
measuring incubation period), and this is achieved with
$\underline{intracerebral}$ (i.c.) injection of a $\underline{high\ dose}$, using a
$\underline{quick}$ agent combination (such as ME7, 79A, or 22C agents
in $\underline{Sinc}^{s7}$ mice) at any age. Under such circumstances a
sequence occurs in which, after a brief interval of a day
or so, agent increases in titre in the lymphoreticular
organs such as spleen (LRS) and reaches a plateau by
around 3-4 weeks. Replication in the CNS commences at
about 10 weeks and the titre continues to rise there until

clinical and eventually terminal disease supervenes
(Figure 1). Under these conditions the programming for
replication in the CNS is evidently initiated at the time
of infection, and CNS replication proceeds independently
of the earlier peripheral replication. This is concluded,
among other reasons (9), from the failure of splenectomy,
either before or after i.c. injections to impede the
disease onset (10). On the other hand, if an intraperi-
-toneal (i.p.) route had been used an incubation period of
around 35 weeks would have resulted and the occurrence and
replication of agent in the CNS would then be dependent
upon prior replication in the LRS (Figure 1). This can
be shown by the profound delay in clinical onset which
follows reduced peripheral agent replication. This can
be achieved by splenectomy either prior to intraperitoneal
injection, which removes replication sites, or following
it, which removes replicating agent as well (10).

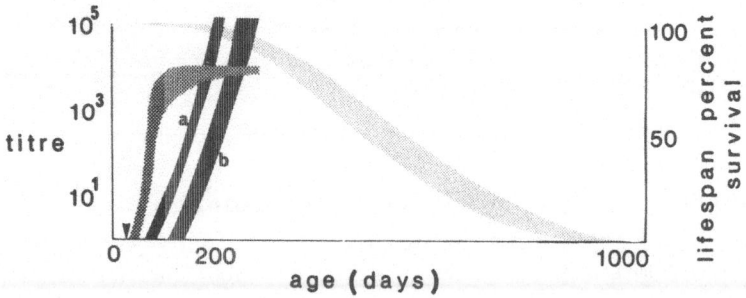

FIGURE 1: Diagram indicating range of titre changes in
spleen and brain during pathogenesis of a "short"
incubation period model of scrapie, following i.c. or i.p.
injection of weanling mice. The bands indicate ranges over
which the dynamics can be controlled with different agents
and hosts. Survival is reduced by the necessary expediency
of sacrificing a proportion of animals prior to natural
death, and this band is therefore not a "lifespan curve".
The entire sequence of events is completed well before
median life span. ▼ - injection time; ▓▓▓ - spleen titre;
▓▓▓ - brain titre; (a - i.c. injection, b - i.p. injection;
░░░ - survival curve.

In contrast, much more prolonged incubation periods follow the use of slow agent combinations (such as 22A in $\underline{Sinc}^{s7}$ mice). Here, by the simple expedient of reducing the dose of initial infection to around 100 LD_{50}'s, there is a prolonged interval, in excess of a year, during which the infectivity is not detected, but after which it rises rapidly in spleen to give a plateau just as in the quicker combinations (Figure 2). These are the types of circumstances in which the full sequence of events involving CNS replication and eventual neurological deterioration do not occur because of the limits imposed by life span (4).

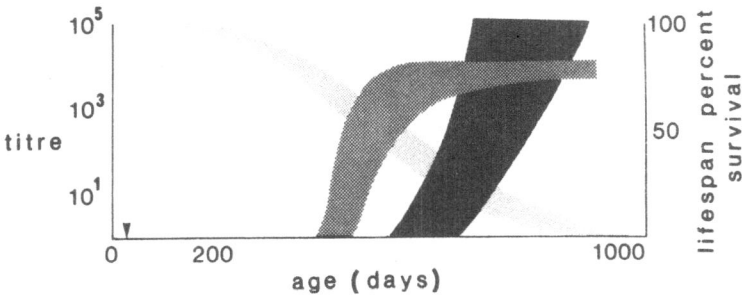

FIGURE 2: Diagram indicating range of titre changes in spleen and brain during pathogenesis of a "long" incubation period model of scrapie, following i.p. injection of weanling mice. With low levels of infection the sequence of events cannot be completed within the life span.

▼ - injection time; ▧ - spleen titre; ▨ - brain titre; ░ - survival curve.

The interval during which there is no detectable
infectivity is provisionally designated as the "zero-phase"
(6), and has also been shown in mice following transmission
from another species (5,11). It is inappropriate at
present to describe this as an "eclipse-phase" because of
a possibility that replication may still be occurring in a
different organ system has not been fully excluded. Also
a failure to detect infectivity may simply reflect the
insensitivity of the in vivo assay. The sequence of
events in pathogenesis, such as the duration of zero-phase,
the rate of replication and the progress of the clinical
disease are programmed by host-controlled steps which are
initiated at the time of initial infection and are not a
consequence of a later randomly-timed triggering of latent
infection.

It has been impossible so far to establish the
identity in the LRS of the systems engaged in the
peripheral pathogenesis of scrapie, although recent work
has cast some light on the biological properties of the
cells involved. Newborn mice (up to 4 days old) are
markedly less susceptible to i.p. infection than are older
mice (12). However by the use of certain immunosuppressive
procedures (steroid treatments) the susceptibility of
older mice to i.p. infection can be reduced to the same
level as that of newborn mice (12, 13, 14).

It has also been shown that immunostimulation
can have the opposite effect (15, 16). However the
possibility that it is not lymphoid components in the
immunological system which are essentially involved in the
peripheral steps in pathogenesis is suggested by the
failure of thymic ablation, either in neonates or in adults
even when combined with irradiation, to effect the disease
in any measurable way (17, 18).

What, then. can be established about the
properties or identity of these cells which are engaged
in the peripheral sequences of scrapie pathogenesis.
Anatomically they are part of the lymphoreticular and
haemopoietic systems (LRS), such as lymph node and spleen

although not prominently in bone marrow (19, 20). A potential list of possible cell types in the LRS having to be considered as primary contenders in peripheral pathogenesis has been made (13). It is important to recognise that, despite the actual replication of agent which is proceeding in the LRS, no pathological or functional deficiency has been found outside the CNS. A possible explanation of this is that any pathological consequence of scrapie replication in spleen is such a low frequency event as to go unrecognised and to be at too low a level to interfere with physiological activity. At the plateau phase in spleen the level of infectivity is such that on average only about one cell in a hundred will carry an infectious unit of scrapie; but whether this is concentrated in only very few cells, or is distributed as lower levels of infection in a larger number of cells, is unknown (6). There is therefore a probability that any pathological events are "diluted out" by the enormous numbers of unaffected cells. In addition the high cell turnover in the LRS may also prevent the occurrence of pathological consequences. Whatever their identity it is clear that the cells involved must represent only a very small proportion of the total LRS population. There is growing evidence that it is a cell - type which is mitotically inactive. Whole body irradiation at lethal or sub-lethal levels is without effect on incubation period (21), and it is this, added to the failure of regeneration of the peripheral replication sites over many weeks following splenectomy even of infant mice (18), that suggests that the cells involved in the peripheral sequences of pathogenesis may be stable cells, as they presumably are also in the CNS.

Most of the biochemical changes found in scrapie, such as the increased DNA synthesis or glyosidase activity, occur late in the course of the disease and are probably secondary to degenerative events (22). However, a biochemical alteration which is early enough to be associated with primary changes, such as agent replication,

is the depression in the levels of polyadenylated RNA
which occurs in the brain and other tissues early in the
incubation period following i.c. or i.p. injections of
139A agent.   In the spleen these levels initially increase
so that at the time when agent titre is rapidly increasing,
at three weeks after injection, the level of polyadenylated
RNA is doubled, to fall later and throughout the remainder
of the disease to depressed levels (23, 24).

        One aspect of the pathogenesis of scrapie remains
a matter of speculation and this concerns the mode of
spread of infectivity in the body, in particular from the
periphery, such as the LRS, to the CNS.   Although it seems
plausible that such access might involve haematogenous
cells, no evidence for this has been found in scrapie
(19, 25, 26);   in experimental C-J disease in guinea pigs
such a mechanism has been suggested, as infectivity was
detected in the buffy coat (27).   However there is now
growing evidence in scrapie in mice that, following
peripheral routes of inoculation, infection might gain
access to the CNS, particularly the spinal cord, via
peripheral nerves (24) as had previously been suggested
in the case of TME in mink by Marsh who has been the main
advocate of this view. (28).

3.   THE PATHOLOGY OF SCRAPIE

The clinical manifestations of scrapie appear to result
from a primary dysfunction of the neuroectoderm, without
primary dysfunction elsewhere.   The lesions which occur
are largely degenerative, without inflammation or evidence
of demyelination.   The lesions are of three main types:
1. a vacuolar degeneration of the neuroparenchyma which
sometimes extends to spongiosis, 2. cerebral amyloid;
3. glial reactions.   It needs to be emphasised that
lesions may be inapparent even at the terminal stage of
the disease, as has been identified in one form of
experimental scrapie in some Cheviot sheep (29);   an almost
identical situation has also been described in TME in aged
Aleutian mink in which the inessential nature of

vacuolation has been shown so unambiguously (30).
Figures 3 and 4 exemplify this lack of histologically
evident degeneration in experimental ovine and murine
scrapie.

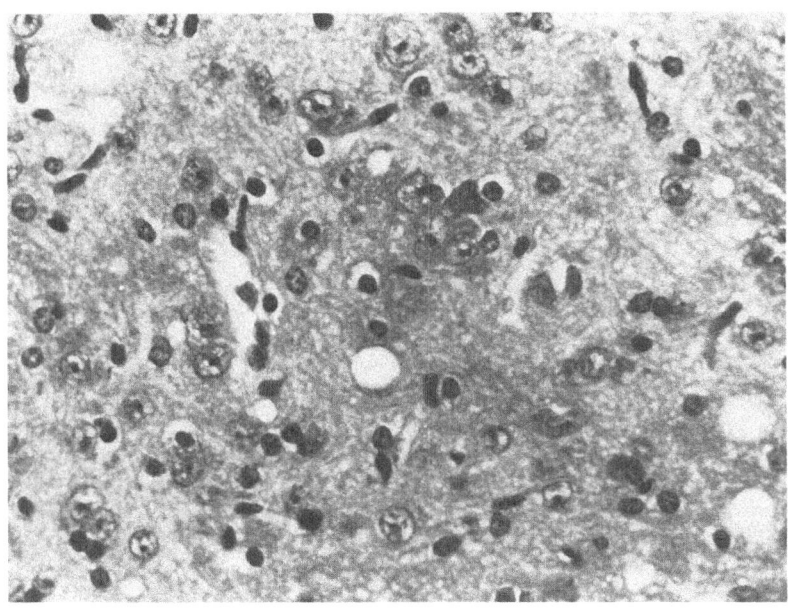

FIGURE 3: Very mild vacuolation in thalamus of a
scrapie-affected RIII mouse, with the '22P' scrapie agent,
which is undergoing identification. No other site in
the brain in routine sections of the whole brain was
more severely affected than this. The incubation
period was 474 days. Haematoxylin and eosin, x 750.

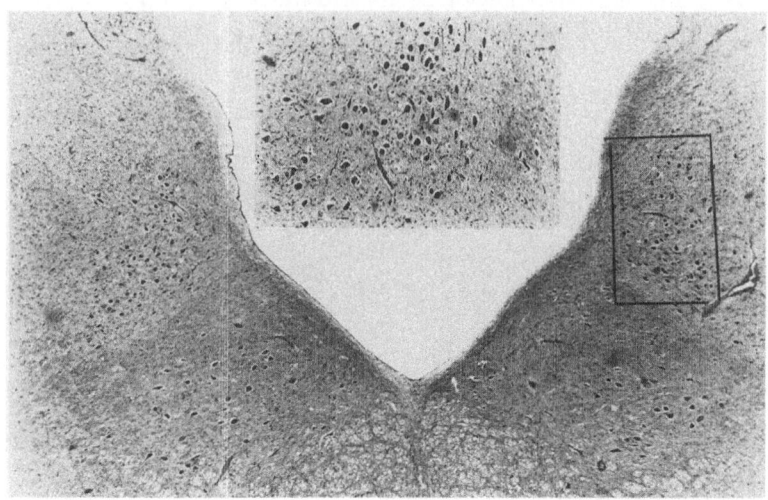

FIGURE 4: Virtual absence of vacuolation in the brain
stem of a scrapie-affected Cheviot sheep with the SSBP/
scrapie source, following subcutaneous injection.   No
other site in the brain in routine sections was more
severely affected than this.   The incubation period was
203 days.   Haematoxylin and eosin, x 40.   Inset shows
the medial vestibular nucleus, x 160.

In many types of scrapie the most prominent lesion is a
degenerative vacuolar change of the grey and sometimes
the white matter.   This is generally diagnostic of
scrapie, but is undoubtedly a secondary morphological
consequence of some unknown primary biochemical defect.
Ultrastructural studies have tended to identify the basis
of this lesion in the body and processes of the neurons,
at least in the grey matter (31, 32, 33).   What are
lacking are systematic ultrastructural studies of lesions
from their inception in different regions of grey and
white matter, in order to identify a putative common
morphological index of a primary molecular pathology.
Any such study would need to include a variety of
different combinations of agent and host, and include

peripheral and intracerebral inoculation methods with
different dose inputs. A major objective of such a study
would be to identify the developmental pathology of early
changes, in grey and white matter, in young, middle-aged
and old animals, and thereby discriminate the early
pathological changes specific for scrapie from the
somewhat similar changes which occur with increasing
intensity as animals age (34, 35, 36). It seems unlikely
that pathological studies on their own can identify the
elusive "cause" or this enigmatic degenerative process,
but carefully conducted studies on the lines suggested
here, concentrating on the early lesions, will narrow the
options for the primary biochemical deficit. Although it
appears likely that a common defect may link the wide
array of different forms and patterns of degeneration in
scrapie and the related encephalopathies, leaving open the
questions of the basis of the variation, a possibility
that different primary defects may occur ought not to be
discounted. For example it has been shown that a similar
depression of choline acetyltransferase occurs in scrapie
and in Alzheimer's disease, but in scrapie this depression
has been found to be much greater in mice affected with
79A scrapie, than with other agent combinations, including
those which represent the closest histopathological
analogues of Alzheimer's disease (1, 37).

It has not been possible so far to identify a
discrete entity in scrapie tissue or homogenates which
might represent the infectious agent itself, although a
variety of different particles, all of unknown
significance, have been found. Several workers have
identified aggregated tubulo-filamentous particles,
frequently in post-synaptic locations and with
approximately similar estimates of size, 20-50nm diameter,
and up to 75nm in length. These have been identified
both in experimental murine scrapie and in the natural
disease in sheep (31, 32, 38, 39 40, 41, 42, 43, 44, 45).
What are probably the same structures have been identified
as spherical particles (21-36nm diameter) in dilated

post-synaptic processes, either as random aggregates, as
vermicular stacks or in overlapping hexagonal patterns in
crystalline array (33).   As these aggregated particles
appear to be the only consistent observation in so many
peoples experience, it is essential that their specific
association with scrapie, and their absence from old
animals and from animals with unrelated neurological
degeneration be confirmed unequivocally by the examination
of properly coded controls.

The vacuolar lesion in grey and white matter at
the level of the optical microscope, shows great diversity
in the patterns of distribution and intensity, and the two
major variables responsible for this variation are the host
genotype and the strain of the agent.   A system for
quantifying these differences has been developed which has
given rise to the establishment of "lesion profiles" whose
major value is that they represent an independent index
for agent identification (21, 29, 46, 47).

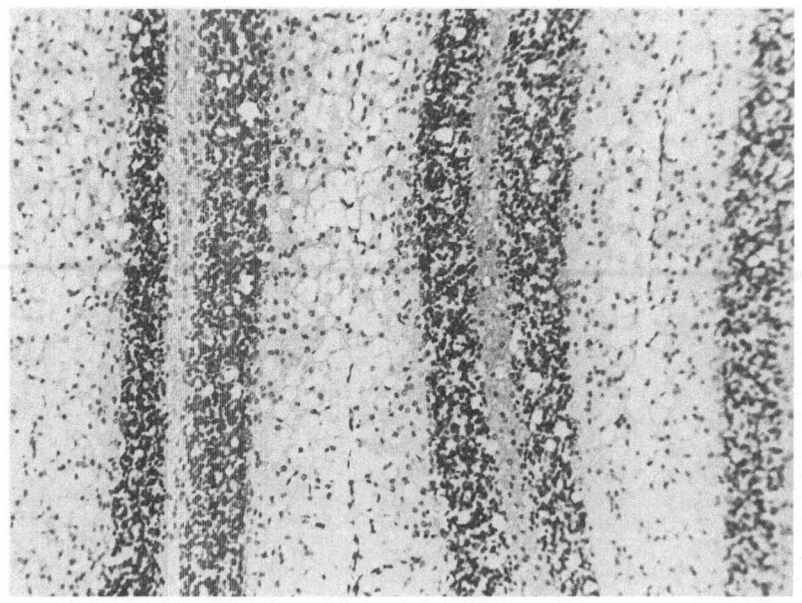

FIGURE 5:   Spongiosis in the cerebellar cortex in an RIII
mouse with 22L scrapie agent. The incubation period was
215 days. Haematoxylin and eosin, x 220.

There are complex interactions between the three
variables of agent strain, host genotype and route of
inoculation, but a lesion profile for any particular
combination is constant over a wide dose range. When all
other factors are held constant, an agent can often be
identified on the basis of its lesion profile calculated
as an average from a group of only 10 mice of a defined
genotype. The vacuolation in different regions of the
brain, both within and between the grey and white matter,
vary independently of one another. However there are
quite big quantitative differences in the overall balance
between grey and white matter involvement between agents.
87V, 87A and 22C represent agents in which the white matter
is virtually unaffected, whereas in 79A and 139A it is
prominently affected. In addition however, mouse strains
differ in their tendency to suffer severe white matter
degeneration with different agents. For example, ME7 is

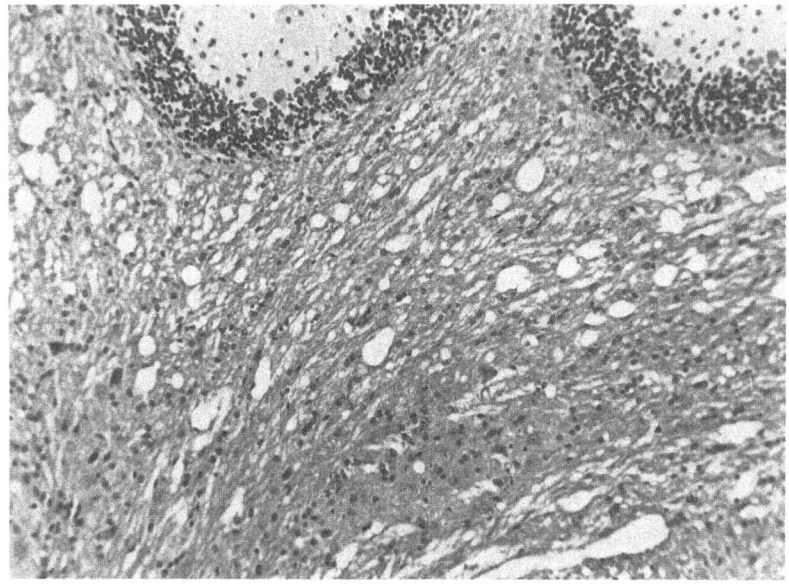

FIGURE 6: Spongiosis in the cerebellar white matter in a
VM mouse with 22L scrapie agent. The cortex is spared.
The incubation period was 224 days. Haematoxylin and eosin,
x 220.

an agent which does not cause this type of pathological
change in most genotypes of mice, but does so in the A2G
and BALB/c strains (29, 48). An analogous situation
occurs with another agent, 22L. A characteristic feature
of 22L is that it induces a widespread and severe
vacuolation, frequently even spongiform, throughout the
molecular and granular layers of the cerebellar cortex, in
all the $\underline{Sinc}^{s7}$ mice which have been tested, and this lesion
is particularly severe in RIII mice (Figure 5). However
in VM mice this lesion is absent, but instead a severe
vacuolar degeneration of white matter occurs which is
absent in $\underline{Sinc}^{s7}$ mice (46)(Figure 6). In a study of
the cerebellar pathology of 22L agent in C57BL's, VM's and
their F_1 crosses, it has been found that the F_1's incurred
intermediate damage (Figure 7). Two important
conclusions drawn from this independence of vacuolation in
white and grey matter is that the white matter change is
not a secondary degeneration resulting from neuronal loss,
and that the lesion in both situations probably reflects
a similar primary defect manifesting itself in different
locations under different circumstances.

As a rule the distribution of vacuolation,
although varying in intensity rostro-caudally, is similar
between the two sides of the CNS, and up to a decade ago
it was widely held that the lesions of scrapie were
bilaterally symmetrical. However, it is now recognised
that some agents give rise to a high frequency of
asymmetrical lesions (Figure 8) (21, 29, 46, 48). These
particular isolates are ones which induce cerebral amyloid
plaques (see below) as a conspicuous feature of their
pathology, although the basis of this association is
unknown. An understanding of asymmetry may be crucial
for elucidating some of the questions surrounding agent
transport and targetting, and lesion distribution
differences. The agents (e.g. 87A) associated with
asymmetrical lesions are also ones which readily break
down, yielding agents (e.g. 7D from 87A) with greatly
altered properties - a phenomenon designated 'Class III

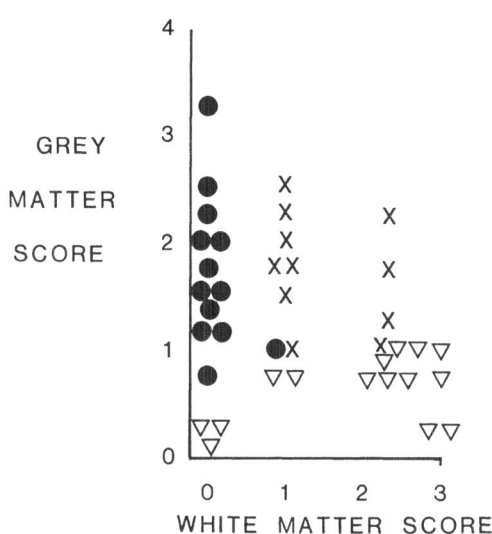

FIGURE 7: Score of spongiosis in individual VM (▽),
C57BL (●) and their F$_1$ cross (X) mice, in grey and white
matter of the cerebellum following i.c. injection of 22L
scrapie agent.   The score in grey matter is the average
of the four scores, on a scale 0-5, in: 1. position 2
(cerebellar cortex adjacent to fourth ventricle) as in
the routine for lesion profiles (8,29,47), and, at the
same transverse levels, the entire : 2. cerebellar cortex,
3. granular layer, 4. molecular layer.   The score in
white matter, on a scale 0-3, as in the routine method
for lesion profiles (21, 29).   All specimens were coded
and scored 'blind'.

stability' (8,49).   One of the most plausible
suggestions explaining the association between asymmetry
and breakdown is that it results from 'mutation', with
loss of some agent coded information, occurring locally
in certain regions of neuroparenchyma thereby yielding
an altered agent (49).   An alternative is that the
asymmetrical foci result from minor and 'faster' component

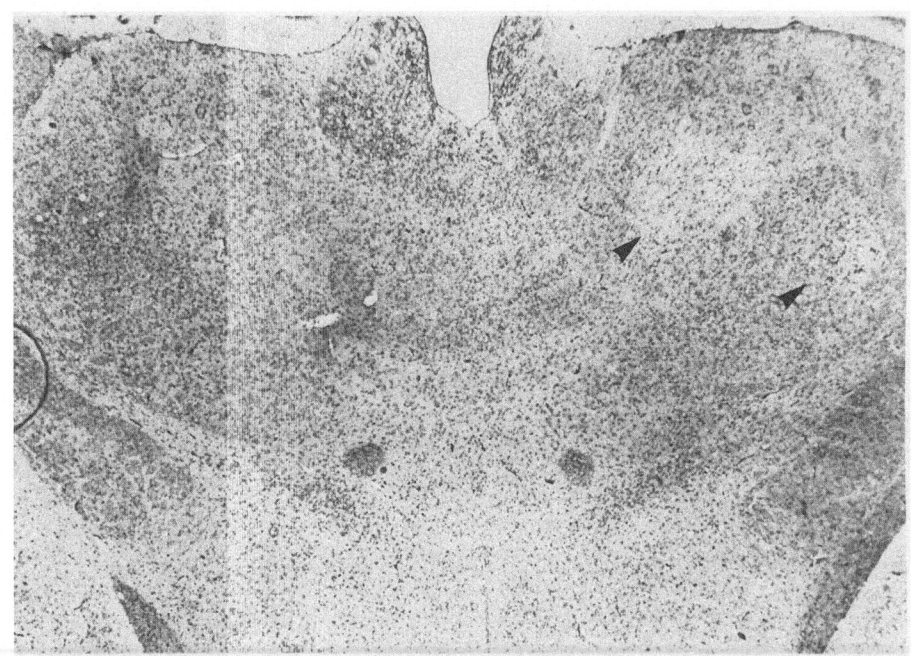

FIGURE 8: Two foci (arrows) of asymmetrical vacuolation
in the thalamus in a VL mouse with 87A scrapie agent. The
incubation period was 444 days. Haematoxylin and eosin, x
50.

in a mixture of agents (21, 46). The slower and major component replicates in restricted regions of the brain such as brain stem and mid-line regions of the tegmentum and hypothalamus, although it is bound with higher priority than the minor component, to sites throughout the rest of the brain. If the major component leaves some sites unoccupied, the minor and potentially faster component has an opportunity to replicate and induce foci of degeneration. In this case it is necessary to postulate a mechanism whereby the relative proportions of both components are maintained in a stable balance throughout subpassage, and a failure of this mechanism would result in the faster component becoming dominant.

It has been further suggested that asymmetrical foci of vacuolation may involve transport of agent along nerve pathways peripherally or centrally (21). This could account for asymmetrical lesions following either i.c. or peripheral injections. Asymmetrical localisation of agent replication could also be a consequence of possible functional asymmetries (50, 51), and this could help to explain the sidedness of asymmetrical lesions following i.c. and peripheral inoculations (46). The agents which give rise to asymmetry, in the forebrain, thalamus and cortices, produce symmetrical lesions in the brain stem and tegmentum. The asymmetrical foci can to some extent to be localised to the side of i.c. injection, and this suggests some localisation of the inoculum. Breakdown might be more likely to occur in regions where the concentration of agent is high, early in pathogenesis, and this provides some rational basis for the finding of asymmetrical lesions in regions of the brain where 7D produces widespread lesions, namely on the mesencephalic tectum, thalamus, hippocampus, and cerebral cortex, all regions in which 87A produces virtually no lesions.

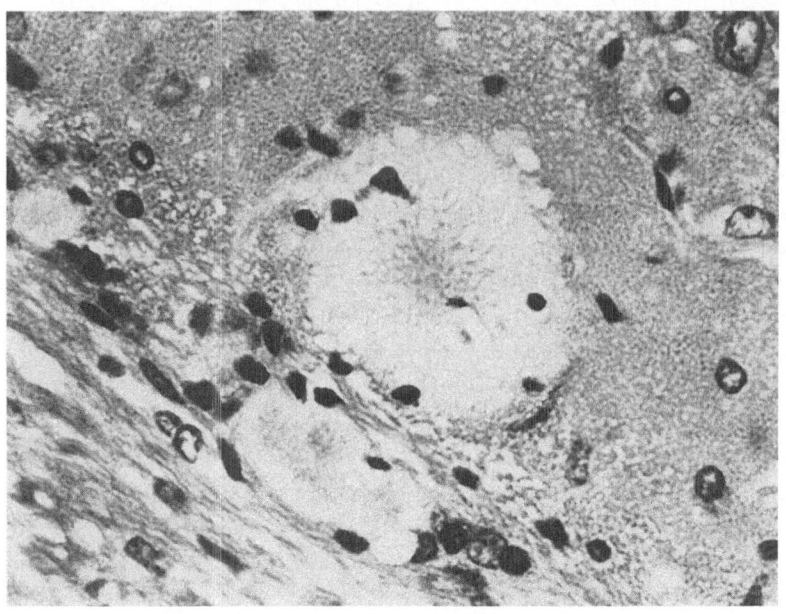

FIGURE 9: Amyloid plaques in the hippocampus (below) and
corpus callosum (above) in a VM mouse with 87V scrapie
agent. The incubation period was 273 days. Haematoxylin
and eosin, x 900.

      Some agents cause severe cerebral <u>amyloidosis</u>
and amyloid plaques in all mice, while with other agents
this lesion is absent. Some agents occupy an intermediate
position, with a moderate plaque incidence in some mouse
genotypes only (52). On the basis of this criterion
some agents can be described as "high" plaque producers
in contrast with most strains which either produce none
or produce a high incidence in some mouse genotypes only
(1). It has been established that there is great variety
in the morphology of the cerebral amyloid in murine
scrapie, and some of the amyloid plaques are closely
analogous to the argyrophilic or senile (neuritic)
plaques of Alzheimer's disease (53, 54) (Figure 9).
Also this lesion can appear indistinguishable from

certain forms of diffuse or perivascular amyloidosis
found in atypical Alzheimer's disease. (55) (Figure 10).
In addition to these examples it has been shown that some
of these forms in murine scrapie are identical to the
amyloid plaques or "kuru-plaques" of kuru and C-J disease
in humans (53, 56). The origin and identity of the
cerebral amyloid in murine scrapie remain unknown, as is
also the case in the human disorders. The most crucial
questions concerning the origin of cerebral amyloid are
firstly whether it depends for its formation on systemic
events with circulating precursors or co-factors, or is
entirely a local neuroparenchymatous phenomenon, and

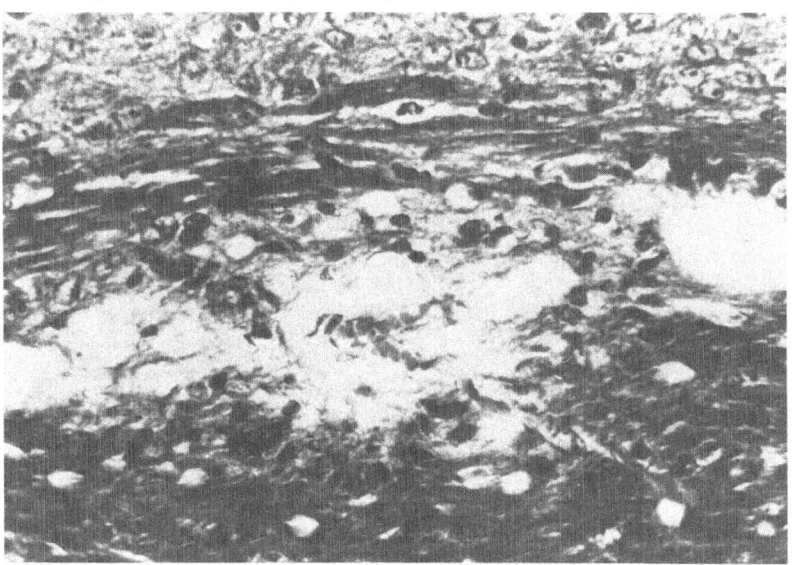

FIGURE 10: Perivascular cerebral amyloid, associated with
small vessels, in the corpus callosum in a VM mouse with
87A scrapie agent, after an incubation period of 613 days.
The amyloid is unstained with phosphotungstic acid
haematoxylin, x 750.

secondly whether it is related to an abortive
immunological response. There are two simple alternatives
for the origin of the amyloid protein (46). It could be
encoded in the genetic information of the agent itself.

The loss of amyloid induction after breakdown i.e. from
87A to 7D (49), might reflect a loss of a region of the
agent code.   The second more likely possibility is that
the cerebral amyloid is a host-coded protein produced in
response to a part of the agent itself or agent coded
products or degeneration products.   In this case the loss
of amyloid induction after break down could be due to a
loss of part of the agent code responsible for the host
stimulus.   In either case the amyloid could have a local
or systemic origin.   The murine scrapie model provides
the only available research tool for an eventual
elucidation of these, and related questions (46).

There is one final aspect of the pathology which
needs to be discussed as it represents a major gap in our
knowledge of the pathology of this disease.   There are

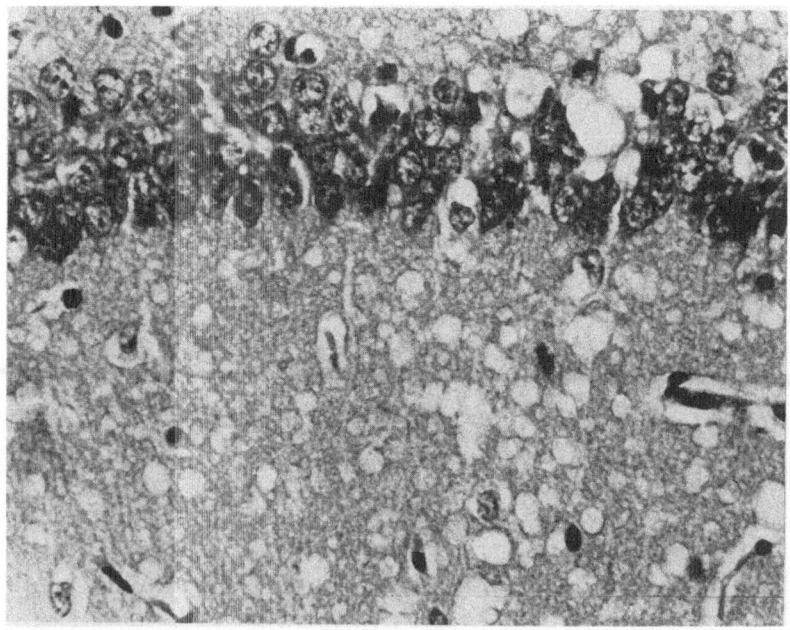

FIGURE 11:   Severe spongiosis, without neuron loss or
glial response, in the hippocampus of a C3H mouse injected
intraperitoneally with ME7 scrapie agent.   The incubation
period was 325 days.   Haematoxylin and eosin, x 750.

numerous unsubstantiated assertions concerning either a
loss or an increase of different cell types of the
neuroectoderm.   The reason that this is an important
area is because it relates to the questions raised earlier
concerning the potential for reversibility of the
degenerative process - if nerve cells die they cannot be
replaced.   With the agents which induce vacuolation in
the forebrain it has been shown that in the hippocampus
a severe vacuolation, with intense spongiosis, without
evidence of a proliferative or infiltrative glial reaction
can be produced following i.p. or other peripheral
inoculation (Figure 11).   However under as yet ill-
-defined circumstances, but only in a small proportion
of cases following intracerebral injection, a complete
loss of the granule cell population of the hippocampus,
accompanied by a profound glial response, microgliosis
and moderate astrocytic proliferation can occur in a small
proportion of cases (Figure 12).

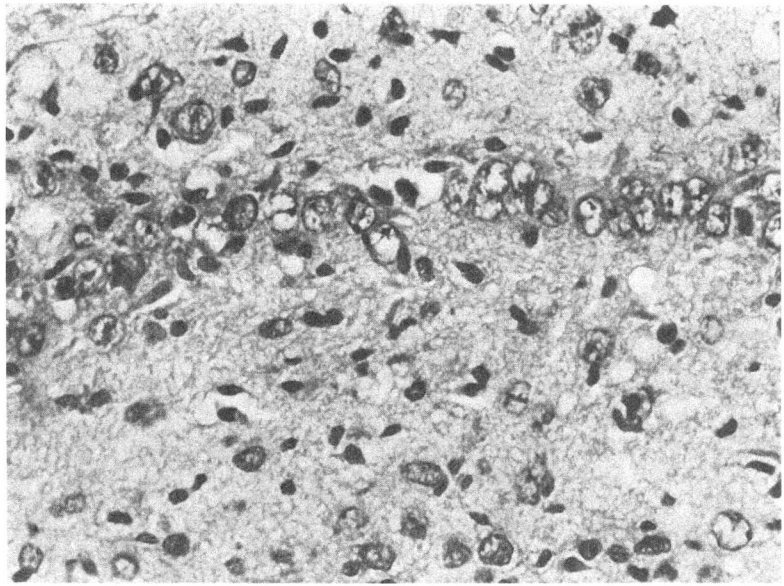

FIGURE 12: Glial response in the hippocampus of a VM mouse
injected intracerebrally with 22A scrapie agent. The
incubation period was 240 days. Haematoxylin and eosin, x
750.

This pattern of degeneration in the hippocampus bears a
striking resemblance to the degenerative pathology in the
hippocampus in a high proportion of cases of Alzheimer's
disease (57).   A similar type of degenerative pathology,
with localised nerve cell loss accompanied by an intense
glial response occurs in the thalamus in a small proportion
of cases following i.c. injection with a number of agents
(21, 24, 46).   An important future objective will be to
distinguish the intermediate situations between the two
extremes, namely the stage when the vacuolar degeneration
still represents a theoretically reversible stage, from the
stage when the degeneration has progressed to irreversible
cell death.   The considerable technical and interpretative
difficulties of defining the quantitative pathology of
scrapie and the related diseases, in terms of putative
neuron loss or astocytosis, have been discussed in detail
elsewhere (21, 46).

4.   CONCLUDING COMMENT

In considering any future pharmacological developments
which may become available there would appear to be two
major objectives: firstly to reverse the deterioration
in the biochemical integrity of the neuroparenchyma before
the degeneration has progressed to an irreversible stage of
cell death;   and secondly to halt the biochemical steps
involved in agent replication in the LRS and CNS.   The
present prospects for thereapeutic advances do not seem
hopeful.   One series of experiments which may offer a
gleam of hope are those involving the site-looking action
of the pathogenesis of a "fast" scrapie agent by the prior
injection of a "slow" agent (6, 8, 58, 59).   These
findings suggest an eventual possibility of mimicking this
site-blocking with non-pathogenic molecules.   However
these experiments have shown that the blocking agent must
undergo replication for blocking to be effective, which
may mean that receptor site turnover occurs, with
resultant elimination of any bound blocking agent.   To be

effective any non-replicating blocking molecule may therefore have to be introducted to the body on a continued basis.

It is important that future work in these areas take account of the wide range of differences in scrapie models, and of the associated different scrapie agents, and not be confined to single and possibly atypical agents or to deviant agents which, during multiple passage in laboratory animals have lost the full expression of their biological capabilities. The clinical and pathological variety in the disease with the various agents makes it highly likely that there will be a wide spectrum of biochemical defects, each with different prominence, and resulting in the different patterns of neurological disturbance.

REFERENCES

1. Dickinson AG, Fraser H, Bruce ME: Animal models for
    dementias. In: Alzheimer's disease, early
    recognition of potentially reversible defects,
    Glen AIM, Whalley LJ (eds), Edinburgh, Churchill-
    Livingstone, 1979 (in press).

2. Kimberlin RH, Walker CA: Characteristics of a short
    incubation model of scrapie in the golden hamster. J
    Gen Virol 34: 295-304, 1977.

3. Dickinson AG: Scrapie in sheep and goats. Chapter
    10 in: Slow virus diseases of animals and man,
    Kimberlin RH (ed), Amsterdam, North-Holland, 1976.

4. Dickinson AG, Fraser H, Outram GW: Scrapie incubation
    time can exceed natural life span. Nature 253: 732
    -733, 1975.

5. Kimberlin RH, Walker CA: Evidence that the transmiss-
    -ion of one source of scrapie agent to hamsters
    involves separation of agent strains from a mixture.
    J Gen Virol 39: 487-496, 1978.

6. Dickinson AG, Outram GW: The scrapie replication site
    hypothesis and its implications for pathogenesis.
    In: Slow transmissible diseases of the nervous
    system, Hadlow WJ, Prusiner SB (eds), Academic Press
    1979 (in press).

7. Dickinson AG, Fraser H: An assessment of the genetics
    of scrapie in sheep and mice. In: Slow
    transmissible diseases of the nervous system, Hadlow
    WJ, Prusiner SB (eds), Academic Press, 1979 (in
    press).

8. Kimberlin RH: this volume, Chapter 2.

9. Outram GW: The pathogenesis of scrapie in mice.
    Chapter 14 in: Slow virus diseases of animals and
    man, Kimberlin RH (ed), Amsterdam, North-Holland,
    1976.

10. Fraser H, Dickinson AG: Pathogenesis of scrapie in
    the mouse: the role of the spleen. Nature 226:
    462-463, 1970.

11. Kimberlin RH, Walker CA: Pathogenesis of scrapie:
    agent multiplication in brain at first and second
    passage of hamster scrapie in mice. J Gen Virol
    42: 107-117, 1979.

12. Outram GW, Dickinson AG, Fraser H: Developmental maturation of susceptibility to scrapie in mice. Nature 241: 536-537, 1973.

13. Outram GW, Dickinson AG, Fraser H: Reduced susceptibility to scrapie in mice after steroid administration. Nature 249: 855-856, 1974.

14. Outram GW, Dickinson AG, Fraser H: Slow encephalo--pathies, inflammatory responses, and arachis oil. Lancet i: 198-200, 1975.

15. Dickinson AG, Fraser H, McConnell I, Outram GW: Mitogenic stimulation of the host enhances susceptibility to scrapie. Nature 272: 54-55, 1978.

16. Kimberlin RH, Cunning**ton** PG: Reduction of scrapie incubation time in mice and hamsters by a single injection of methanol extracted residue of BCG. FEBS Microbiol Lett 3: 169-172, 1978.

17. McFarlin DE, Raff MC, Simpson E, Nehlsen SH: Scrapie in immunologically deficient mice. Nature 233: 336, 1971.

18. Fraser H, Dickinson AG: Studies of the lymphoreticular system in the pathogenesis of scrapie: the role of the spleen and thymus. J comp Pathol 88: 563-573, 1978.

19. Eklund CM, Kennedy RC, Hadlow WJ: Pathogenesis of scrapie virus infection in the mouse. J Infect Dis 117: 15-22, 1967.

20. Dickinson AG, Fraser H: Genetical control of the concentration of ME7 scrapie agent in mouse spleen. J Comp Pathol 79: 363-366, 1969.

21. Fraser H: Scrapie: a transmissible degenerative CNS disease. In: Progress in Neurological Research, Behan PO, Clifford Rose F (eds), London, Pitman, 1979 (in press).

22. Kimberlin RH: Experimental scrapie in the mouse: a review of an important model disease. Science Progress, Oxford 63: 461-481. 1976.

23. Corp CR, Kimberlin RH: Changes in the metabolism of **poly** adenylated RNA commencing early in the incubation period of scrapie in the mouse. Vet Microbiol 2: 193-204, 1977.

24. Kimberlin RH: Early events in the pathogenesis of scrapie in mice: biological and biochemical studies. In: Slow transmissible diseases of the nervous system, Hadlow WJ, Prusiner SB (eds) Academic Press, 1979 (in press).

25. Dickinson AG, Meikle VMH, Fraser H: Genetical control
    of the concentration of ME7 scrapie agent in the
    brain of mice.   J Comp Pathol 79:   15-22, 1969.

26. Clarke MC, Haig DA: Presence of the transmissible
    agent of scrapie in the serum of affected mice and
    rats.   The Vet Rec 80:   504, 1967.

27. Manuelides EE, Gorgacz EJ, Manuelides L: Viraemia in
    experimental Creutzfeldt-Jakob disease.   Science
    200:   1069-1071, 1978.

28. Marsh RF, Miller JM, Harrison RP: Transmissible mink
    encephalopathy: studies on the peripheral
    lymphocyte.   Infect Immun 7:   352-355, 1973.

29. Fraser H: The pathology of natural and experimental
    scrapie.   Chapter 12 in: Slow virus diseases of
    animals and man, Kimberlin RH (ed), Amsterdam,
    North-Holland, 1976.

30. Marsh RF, Sipe JC, Morse SS, Hanson RP:
    Transmissible mink encephalopathy, reduced spongi-
    -form degeneration in aged mink of the Chediak-
    Higashi genotype.   Lab Invest 34:   381-386, 1976.

31. Lampert P, Hooks J, Gibbs CJ, Gajdusek DC:
    Altered plasma membranes in experimental scrapie.
    Acta Neuropathol (Berl) 19:   81-93, 1971.

32. Lampert P, Gajdusek DC, Gibbs CJ: Subacute
    spongiform virus encephalopathies.   Scrapie, Kuru
    and Creutzfeldt-Jakob disease: a review.   Am
    J Pathol 68:   626-646, 1972.

33. Barringer JR, Prusiner SB: Experimental scrapie in
    mice, ultrastructural observations.   Ann Neurol
    4:   205-211, 1978.

34. Field EJ: The significance of astroglial
    hypertrophy in scrapie, kuru, multiple sclerosis
    and old age together with a note on the possible
    nature of the scrapie agent.   Dtsche Z Nervenheilk
    192:   265-274, 1967.

35. Fraser H: Comparative morphology of ageing and
    scrapie.   In: Proceedings of the VIth
    International Congress of Neuropathology: p.897,
    Paris, Masson and Cie, 1970.

36. Gajdusek DC: Slow virus infection and activation of
    latent infections in ageing.   Adv Gerontol Res 4:
    201-218, 1972.

37. McDermott JR, Fraser H, Dickinson AG: Reduced choline
    acetyltransferase activity in scrapie mouse brain.
    Lancet ii: 318-319, 1978.

38. David-Ferrieira JF, David-Ferrieira KL, Gibss CJ, Morris JA: Scrapie in mice: ultrastructural observations in the cerebral cortex. Proc Soc Exp Biol Med 127: 313-320, 1968.

39. Bignami A, Parry HB: Aggregations of 35-nanometer particles associated with neuronal cytopathic changes in natural scrapie. Science 117: 389-390, 1971.

40. Bignami A, Parry HB: Electron microscopic studies of the brain of sheep with natural scrapie. Brain 95: 319-326.

41. Narang HK: Virus-like particles in natural scrapie of sheep. Res Vet Sci 14: 108-110, 1973.

42. Narang HK: Ruthenium red and lanthanum nitrate, a possible tracer and negative stain for scrapie "particles". Acta Neuropathol (Berl) 29: 37-43, 1974.

43. Narang HK: An electron microscopic study of the scrapie mouse and rat: further observations on virus-like particles with ruthenium red and lanthanum nitrate as a possible trace and negative stain. Neurobiology 4: 349-363, 1974.

44. Lamar CH, Gustafson DP, Krusovich M, Hinsman EJ: Ultra structural studies of spleen, brains and brain cell culture of mice with scrapie. Vet Pathol 11: 13-19.

45. ZnRhein GM, Varakis I: cited by Marsh RF: The subacute spongiform encephalopathies. Chapter 15 in: Slow virus diseases of animals and man, Kimberlin RH (ed), Amsterdam, North-Holland, 1976.

46. Fraser H: Neuropathology of scrapie: the precision of lesions and their diversity. In: Slow transmissible diseases of the nervous system, Hadlow WJ, Prusiner SB (eds) Academic Press, 1979 (in press).

47. Fraser H, Dickinson AG: Agent-strain differences in the vacuolation and intensity of grey matter vacuolation. J Comp Pathol 83: 29-40, 1973,

48. Fraser H, Bruce ME, Dickinson AG: Quantitative pathology for understanding the nature and pathogenesis of scrapie, using different strains of agent. In: Proceedings of the VIII the International Congress of Neuropathology: p.277, Amsterdam, Excerpta Medica, 1975.

49. Bruce ME, Dickinson AG: Biological stability of different classes of scrapie agent. In: Slow transmissible diseases of the nervous system, Hadlow WJ, Prusiner SB (eds) Academic Press, 1979 (in press).

50. Galaburda AM, Le May M, Kemper TL, Geschwind N,: Right-left asymmetries in the brain. Science 199: 852-856, 1978.

51. Kinsbourne M: Asymmetrical function of the brain. Cambridge and London, Cambridge University Press, 1978.

52. Bruce ME, Dickinson AG, Fraser H: Cerebral amyloidosis in scrapie in the mouse: effect of agent strain and mouse genotype. Neuropathology and Applied Neurobiology 2: 471-478, 1976.

53. Fraser H, Bruce ME: Argyrophilic plaques in mice inoculated with scrapie from particular sources. Lancet i: 617, 1973.

54. Wisniewski HM, Bruce ME, Fraser H: Infectious etiology of neuritic (Senile) plaques in mice. Science 190: 1108-1110,1975.

55. Corsellis JAN, Brierley JB: An unusual type of pre-senile dementia (atypical Alzheimer's disease with amyloid change). Brain 77: 571-587, 1954.

56. Bruce ME, Fraser H: Amyloid plaques in the brains of mice infected with scrapie: morphological variation and staining properties. Neuropathology and Applied Neurobiology 1: 189-202,1975.

57. Corsellis JAN: The limbic areas in Alzheimer's disease and in other conditions associated with dementia. Chapter In: Alzheimer's disease, a Ciba Foundation Symposium, Wolstenholme GEW, O'Connor M (eds) London, Churchill, 1970, p37-45.

58. Dickinson AG, Fraser H, Meikle VMH, Outram GW: Competition between different scrapie agents in mice. Nature, New Biology 237: 244-245, 1972.

59. Dickinson AG, Fraser H, McConnell I, Outram GW, Sales DI, Taylor DM: Extraneural competition between different scrapie agents leading to loss of infectivity. Nature 253: 556, 1975.

DISCUSSIONS OF PAPERS BY R.H. KIMBERLIN AND H. FRASER

This began with a consideration of the mode of spread of scrapie in vivo. With the exception of one report of the presence of Creutzfeldt-Jacob disease agent in the buffy coat of guinea-pigs shortly after inoculation, there was no evidence of the presence of this type of agent in the peripheral blood. It was later speculated that infection of a small subpopulation of lymphocytes (e.g. B lymphocytes) might explain some of the features of scrapie; but hard facts were not available. Infection with scrapie probably spread from spleen to brain; splenectomy apparently removed replication sites but there were others in the reticulo-endothelial system and the zero phase was not prolonged. In tissue culture experiments, using the SMB cell line (persistently infected with scrapie), replication appeared to be associated with membranes and endoplasmic reticulum.

A question was asked concerning the role of interferon and whether it influenced infectivity; the answer was, broadly, no. There was no relation between the production of amyloid in scrapie-infected mice and the depression of choline acetyl transferase activity; strains which produced such a depression did not necessarily produce amyloid and there was no correlation between a long incubation period and a high incidence of amyloid. It was not known whether scrapie amyloid contained Ig-like chains. Reports of an inhibition of immune response in scrapied animals have been made but they are only just statistically significant and limited to one system; consequently, it seemed more likely that this phenomenon was non-specific.

Vertical transmission had not been detected and, indeed, immunological competence was necessary for scrapie infection. Day old mice were less readily infected, or the incubation period was longer than weaned mice but, otherwise, age dependence was minimal. This neonatal immunological incompetence was an important difference between sheep and mice. Kuru was not transmitted vertically but (Dr Cathala) the situation might be different in Creutzfeldt-Jacob disease.

Professor Sanger drew attention to the alleged resemblance of the alterations of the plasmalemma (plasmalemmesomes) in plants with viroid infection and their resemblance to similar changes in scrapie

(vacuoles). Such a comparison was fallacious as plasmalemmasomes are
seen also in quite healthy plants.

The nature of scrapie and its relation to kuru and to Creutzfeldt-
Jacob disease were discussed at length. Although the pathology of
these agents differed in mice, the basic vacuolation, gliosis and
plaques remained. Trans-species transmission was sometimes possible
but not invariably so (e.g. the failure to transmit transmissible mink
encephalopathy in mice). The unusual nature of the scrapie agent was
highlighted by its unusual response to UV irradiation. Dr Kimberlin
felt that scrapie nucleic acid used host components which fitted
(much as a hermit crab) and, together these components formed a fully
infectious agent. It seemed much more likely that scrapie was an
unusual virus than a viroid and Dr Kimberlin closed by quoting
Dr Dickinson who had coined the neologism 'Virelo' for agents such
as scrapie.

SUBACUTE SCLEROSING PANENCEPHALITIS: CHARACTERIZATION OF THE ETIOLOGICAL AGENT AND ITS RELATIONSHIP TO THE MORBILLI VIRUSES

R. STEPHENSON AND V. TER MEULEN

1. INTRODUCTION

Subacute sclerosing panencephalitis (SSPE) is a slow virus infection of the central nervous system (CNS) affecting children and young adults (1,2). The clinical picture reflects a variable and wide spread involvement of the CNS. The disease usually starts with mental and behavioural changes, which after a period of weeks or months are followed by characteristic neurological symptoms. They are characterized by various disturbances of motor function, myoclonic jerks and epileptic seizures. At a later stage of the disease progressive cerebral degeneration with symptoms and signs of decerebration occurs leading always to death. Neuropathologically the disease process is characterized by perivascular cuffing, increase in hypertrophic astrocytes, proliferation of microglia as well as demyelination and the presence of intranuclear Cowdry type A and B inclusion bodies. This latter finding, which was already described in 1933 (3) has always been interpreted as footprints of a viral infection, but the first direct evidence of virus involvement was reported in 1965 by Bouteille et al. (4) who observed tubular structures resembling paramyxovirus nucleocapsids in brain tissue of SSPE patients. These structures were soon identified as measles virus nucleocapsids as infectious measles virus (referred to as SSPE virus) was isolated from biopsy material. Although these findings fulfil Koch's first postulate namely the occurrence of the 'parasite in every case of the disease in question' and incriminates measles virus as the causative agent in this disease, the pathogenesis is still not understood. If measles virus is involved, then additional factors,

either host or virus derived, must play a pathogenetic
role, since rarity and rural prevalence of this disease
cannot be correlated to a ubiquituous measles infection.

This paper summarized the virological findings in
SSPE from recent investigations, compares the SSPE virus
characteristics with those of related viruses and dis-
cusses the possible pathogenetic mechanisms underlying
this slow virus infection.

2. VIROLOGICAL AND IMMUNOLOGICAL EVIDENCE FOR MEASLES VIRUS INFECTION IN SSPE

2.1. *Presence of measles virus antigen and nucleocapsids in brain cells*

Neurological investigations have shown that in diseased
brain areas of SSPE patients intranuclear and intracyto -
plasmic eosinophilic inclusion bodies occur in neurons,
astrocytes and oligodendroglia cells (1). These inclusion bodies contain RNA by cytochemical staining and
consists of paramyxovirus-like nucleocapsids of diameter 17 - 23 nm and of length up to 500 nm. The nucleocapsids are found in both nucleus and cytoplasm of
brain tissue culture cells, and also in cells persistently or lytically infected with virus isolated from SSPE
patients. The nuclear nucleocapsids have always a
'smooth' appearance and those in the cytoplasm a rough
appearance. This distinction between nuclear and cytoplasmic morphology is constant whether autopsy material
is observed or that from lytically or persistently infected cells is examined. Only the rough nucleocapsids
appear in released virus, whatever the source. By immunofluorescent staining these intranuclear inclusions
clearly stain with measles antibodies. Moreover measles
antigen can be detected in cell processes, suggesting the
spread of viral material along this route.

2.2. Isolation of SSPE virus

All standard methods for the isolation of SSPE virus
from SSPE brain material failed to recover infectious
virus. Only after cocultivation of SSPE brain tissue
culture with cultures susceptible to measles virus

replication was SSPE virus isolated. However, it is
important to emphasize that the isolation of SSPE virus
is an exception rather than the rule (5). Virus has been
recovered not only from biopsy and necropsy material,
but also in one instance from lymph nodes (6). The characteristics of these virus isolates vary from a virus
which produces c.p.e. and free infectious virus after
only a few subcultures to those who need many subcultures for complete expression. Other isolates may establish a persistent infection with no free virus produced
and mainly subgenomic nucleocapsids and viral antigens
being detected (7,8).

2.3. Immuneresponse to measles virus

One of the main diagnostic criteria in SSPE is the presence of high titre measles antibodies in serum and C.S.F.
specimens. All patients exhibit a strong hyperimmunere-
sponse against all biological active measles virus antigens. Measles specific IgG antibodies represent 10 % to
20 % of the total serum IgG and about 75 % of the total
C.S.F. IgG (9). Moreover, the antibodies in the C.S.F.
are of oligoclonal nature and reveal restricted variabi-
lity of the V-region of Kappa and Lambda light chains
and restricted specificities. This finding together with
a low ratio of serum to C.S.F. antibodies suggest a local production of measles antibodies by invading lymphocyte clones. These antibodies neutralize infectious
measles virus, lyse measles infected target cells in the
presence of complement and induce antigen capping in cell
lines persistently infected with measles virus. However,
our analysis of antibody activities against all measles
virus structural proteins by immunprecipitation in combination with P.A.G.E. revealed that the hyperimmune
reaction in SSPE is only directed against structural proteins H, P, N and F and not against the M or L proteins
(Fig.1). Similar findings are obtained with convalescent
sera from cases of simple of complicated measles.
Probably these two proteins are either poor immunogens,
not produced in sufficient quantities or not released
from infected cells to be available for the immune

64

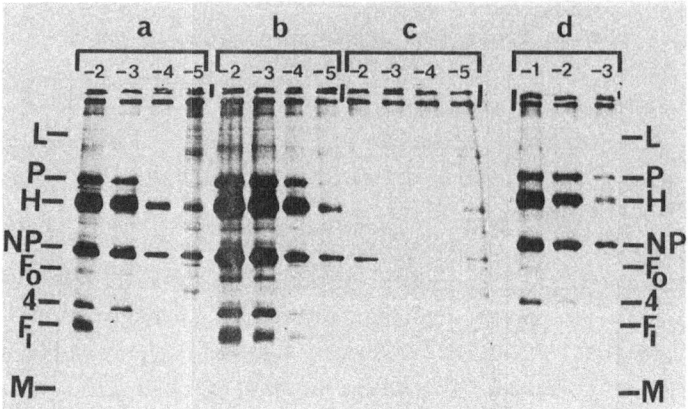

Figure 1

Immune precipitation of measles-specific polypeptides by
sera from human patients. Vero cells were infected with the
LEC strain of SSPE virus and labelled with ^{35}S -methionine.
Cells were lysed in buffer containing 0.15 M NaCl, 10 mM
Tris pH 7.4, 1 mM EDTA, 1 % NP40, 0.1 % Na Azide, 2 mg/ml
PMSF and 1,000 units/ml Aprotinin. Cells were frozen,
thawed, sonicated and clarified at 100,000 g for 30 min.
200 µl of lysate were reacted with 5 µl of sera at the appropriate dilution. The precipitates were incubated with
fixed Staphylococcus aureus bacteria, washed and boiled in
sample buffer (5M Urea, 1 % SDS, 0.1 % BME). Samples were
run on 15 % discontinuous gels containing 1 % SDS and 5 M
Urea. Polypeptides designated as described previously (19).
a) Sera from a classical SSPE patient
b) Sera from a patient with viral meningitis
c) Sera from a 6 months old infant with no known measles
   infection or contact.
d) Convalescent sera 5 months after acute measles.
Tracks are designated by the indice of sera concentration.

competent cells.

In contrast to this humoral hyperimmune reaction no generalized defect of cell-mediated immunity (CMI) has been found in SSPE patients except for a specific unresponsiveness to measles virus on skin testing (2). Inoculation of measles virus antigen does not lead to a positive skin reaction which has been interpreted as a possible counterpart to the transient tuberculin anergy observed in measles disease. Otherwise blastogeneic and lymphokine responses are unimpaired after stimulation with mitogens or antigens unrelated to measles virus. Moreover, skin allografts are rejected in a normal fashion. However, controversial results are reported by using measles virus antigen in assays for CMI. Both normal and abnormal responses have been repeatedly found, which may indicate that a measles specific CMI defect could exist in SSPE (2). However, until the functional state of T lymphocyte in SSPE is defined, no definite answer to the role of the immune response in relation to this chronic disease process can be given (11).

3. PHYSICO-CHEMICAL PROPERTIES OF MEASLES VIRUS

3.1. *Morphology*

Measles virus is a member of the morbillivirus group which include canine distemper (CDV), rinderpest virus (RV) and pest des petits ruminants virus (PRV) and form a subgroup of the genus Paramyxoviridae (ICNV cytptogram: R/1 : 4-8/1 : S/E : v/o) (12). These viruses have a pleomorphic structure with a diameter of 100 - 300 nm. The outer layer of these viruses consists of a lipid bilayer from which protrude spikes of 9 - 15 nm in length. This heterogeneity in the size of the spikes may be due to more than one envelope protein being present. The lipid bilayer surrounds a coiled hollow nucleocapsid similar in appearance to that found in other paramyxoviruses (13).

Reported values for the buoyant density of the purified virus particle in either CsCl or sucrose vary between 1.24 and 1.22 gm/cc for infectious virus, and drops to 1.20 gm/cc for virus with a reduced infectivity

(14,15). This variation in density and the lack of any
data on sedimentation values is probably due to significant and variable amounts of host protein in the virus
preparations analysed. Also the presence of a fusion protein in the outer membrane of the virus results in decreased rigidity of the viral envelope and makes biophy-
sical examination difficult. Nucleocapsids isolated from
purified virus have a density of 1.30 - 1.31 gm/cc and sediment between 100 and 200 S. When examined by electron
microscopy these nucleocapsids are hollow, have an exter-
nal diameter of 15 - 20 nm, an internal diameter of 7 nm,
a pitch of 6.6 nm and lengths up to 1,000 nm, although
shorter species predominate: rarely circular and branched
forms can be seen. Nucleocapsids from purified measles
virus and the cytoplasm of infected cells have a 'rough'
appearance when examined by electron microscopy, whereas
those from the nucleus have a 'smooth' texture. This sub-
cellular difference in nucleocapsid morphology is invari-
ent whether the specimen comes from lytically infected
cells, persistently infected cells or from fresh autopsy
material (8,16). The nucleocapsid contains about 5 % by
weight of RNA. The RNA from the purified virus has a minor component sedimenting at 50S and a major component
sedimenting from 12 - 30S (17).

So far isolates of SSPE virus show no gross morphological differences from that described above for measles
virus except that in both persistent and lytic infections
with SSPE, the nuclear form of the nucleocapsid predomi-
nates, whereas with measles it is the cytoplasmic form
which is most common (2).

3.2. Molecular structure of the virion

Although there is some discussion in the literature as to
the precise number and molecular weight of morbillivirus
polypeptides, the genome of this group of viruses appears
to specify six primary gene products. The molecular
weights of these proteins show some similarity among different members of the group but are not identical in every case (15,18). The largest polypeptide has an apparent
molecular weight of between 100 - 150 K. There are two

glycoproteins which by analogy to Sendai virus are assumed
to be the Haemagglutinin (HA) and the haemolysin or fusion
(F) protein (19). However, the corresponding biological activities have been reported only for measles virus (MV) in
spite of many attempts to detect them in canine distemper
and Rinderpest. The haemagglutinin of these viruses has a
molecular weight of about 80,000 daltons. The fusion protein is found as a glycosylated precursor of molecular
weight 60,000 in infected cells but is only found in the
cleaved form in the virus. The cleavage products have molecular weights of 40,000 and 18,000 and only the latter
is glycosylated (19,20). The nucleocapsid contains two
proteins, a minor species of 70,000 daltons and a major
species of 62,000 daltons. The smallest major virion polypeptide has a molecular weight of 34,000 to 37,000 daltons and from its structural similarities to Sendai virus
M protein, is assumed to be the protein lining the inner
membrane of the virion (21). A further polypeptide of
molecular weight 55 K has been reported in MV infected
cells (18,19). Its function has not yet been elucidated
although a similar protein in Sendai virus has been shown
to be derived from the major nucleocapsid protein (22).

SSPE viruses reveal, in general, the same properties.
However, by comparison of the RNAs of measles and SSPE
viruses by hybridization, some differences were noted.
One study showed that SSPE 50S RNA shared 60 % homology
with measles 18S RNA, although no attempt was made to determine the role played by negative stranded subgenomic
defective RNAs (23). A more detailed analysis using oligo
dT isolated mRNA and competition hybridization suggested
that SSPE mRNA had a higher genetic complexity than
measles mRNA (24). These authors have also reported differences in migration on SDS P.A.G.E. of at least one
mRNA from SSPE and measles viruses (25). Similar small
differences in apparent molecular weight have been reported for the M and P proteins for several strains of
SSPE (26). However such variation also appears to occur
between measles isolates as well (27). Recently the 50S
RNA of 2 measles and one SSPE isolate from the cytoplasm

of AMD-treated infected cells have been compared by analysis of the oligonucleotides from a T_1 digest (28).
These data show considerable similarity but not identity
between the three isolates. However, at present it is
not possible to show that any differences are specific
for SSPE virus.

4. BIOLOGICAL PROPERTIES

4.1. *Comparative Antigenicity*

4.1.1. Antigenic relationships between measles and SSPE virus

In general, little difference has been found in the
antigenicity of these viruses when assayed by neutralisation, haemolysin inhibition, haemagglutinin
inhibition (2) and indirect radio-immunoassay (21).
Moreover, all the virus-specific polypeptides present in cells infected with an SSPE virus can be
precipitated by antisera against measles virus
(Fig. 2). However, when antigenic studies were
carried out on purified M protein from one SSPE and
one Measles isolate, they revealed antigenic differences between the M proteins of these two viruses
(25). The immunological differentiation between
these proteins was achieved by using rabbit sera
after a short-term immunization with purified protein.

4.1.2. Antigenic relationships between measles virus and canine distemper virus.

It has been known for some time that human measles
convalescent sera cross-react with CDV (29). However, sera taken during the acute phase of MV infection showed no heterologous activity. Experiments on animals infected with measles virus gave
similar results (30. Sato and co-workers (31) reported that sera from SSPE patients cross-react to a
much higher degree with CDV than do either patients
with acute atypical measles, convalescent vaccinees,
or hyperimmune animals. However, it is difficult to

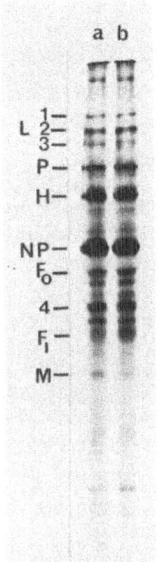

Figure 2

Immune precipitation by rabbit sera of virus-specific polypeptides from cells infected with SSPE virus (LEC isolate). Immune precipitation and electrophoresis of radio-labelled cell lysates were performed as in Figure 1.

a) Sera raised against Measles virus (Edmonston)

b) Sera raised against SSPE virus (LEC)

Polypeptides designated as for Figure 1.

---

compare different studies in a quantitative fashion as sera of widely varying titres are used. Data on the reverse process, i.e. the MV neutralization activity of CDV antisera is equivocal. In general, CDV antiserum contains little or no neutralizing activity to measles. The above studies on virus neutralization presumably only detect surface antigens responsible for attachment and penetration; other studies on infected cells, involving immune fluorescence (32) and electron microscopy of ferritin tagged antibody show a much higher degree of cross-reactivity, presumably because reactions with many of the internal antigens are detected. Similar high levels of cross-reactivity have been observed using radio-immune assay directed against whole disrupted virions (21). Studies on antigenic similarities between individual virus components reveal that the nucleocapsid of MV and CDV are identical antigenically and that sera raised against CDV have a high level of antibody against MV Haemolysin and low activities against MV Haemagglutinin (33).

4.2. Host range and growth kinetics

Although several reports are available which show differences in tissue culture, host range and growth kinetics between measles virus and SSPE virus, no parameter has been found to distinguish SSPE virus from measles

virus per se. It has been observed that some SSPE isolates
have a wider, others a narrower host range, when compared
with wild-type measles virus (34). Moreover, primary human brain cultures can easily be infected with measles
virus but not with SSPE virus (34,35). In similar studies
using hamster dorsal root ganglia one of three measles isolates and one of two SSPE isolates were infectious (36).
Other workers have compared virulence, HA production and
intracellular antigens by FA staining of SSPE and measles
isolates, and found that, whereas the SSPE isolates differed from each other, they each shared different properties withMeasles virus (37). In addition, the growth kinetics of two SSPE isolates differed from those of two wild-
type measles isolates (35) and in both cases, the SSPE viruses grew slower than did the measles isolates. However,
it must be emphasized that the passage history and isolation procedure of each isolate under test is seldom
comparable, and thus the relative degree of adaptation
for each isolate is not known.

5. PATHOGENETIC ASPECTS

The virological findings of measles virus structures
and antigens in SSPE brain material as well as the humoral hyperimmune response of these patients to measles
virus has led to many hypotheses in the attempts to explain etiology and pathogenesis of this CNS disorder
(38). Based on epidemiological observations of a preceeding acute measles infection years before onset of
SSPE it is assumed that at this stage measles virus
enter the CNS and a persistent infection is established.
This event probably marks the beginning of this slow
virus infection. However, before a clinically recogniz-
able disease develops this infection is silent for an
average of 5 - 6 years. No direct and definite information is available about the virus host relationship
during this incubation period and the mechanisms which
activate virus latency. Under normal conditions, infectious measles virus entering a sero negative host

induces acute measles which can be overcome by the host
defense mechanisms, if the cellular immune system (CMI) is
unimpaired. Clinical experiences have revealed that individuals with congenital, acquired or iatrogenic deficiency
of thymus dependent lymphocytes usually develop severe
often fatal acute measles whereas children with hypogammaglobulinaemia develop normal measles without complications
and acquire lifelong immunity (39). Based on these observations Burnet suggested that SSPE is the consequence of
a measles specific T cell defect which prevents successful elimination of this virus and therefore allows the
virus to establish persistency in the host (40). So far,
no major immune defect has been found in SSPE which would
explain the rarity, rural prevalence and pathogenesis of
this disease or would support Burnet's hypothesis. Obviously the virus-host relationship underlying this disease
process plays a major role in SSPE and from a virological
point of view the mechanism of measles virus persistency
in brain cells has to be explained. Virological studies
have shown that isolation procedures to rescue infectious
measles virus from SSPE brain material are only successful if infected brain cells are cocultivated with tissue
culture cells susceptible to measles virus replication,
since infectious virus cannot be found directly (2). The
observation of a state of measles virus defectiveness in
brain cells is supported by in vitro studies of persistently infected SSPE brain tissue cultures. The biological and biochemical characterization of these tissue
culture cells revealed the presence of measles antigen,
nucleocapsids and salt dependent haemagglutin in the
absence of infectious virus. All attempts to initiate reactivation of virus synthesis were unsuccessful (7,8). In
addition, biochemical analysis of intracellular RNA demonstrated that the majority of the virus RNA produced was
defective or subgenomic and that the 50S RNA normally
associated with infectious virus was only detectable in
small amounts. These findings indicate an impaired state
of virus replication which may represent the one responsible for virus persistency in SSPE brain. By what

mechanisms such persistency is induced and maintained is
not known, but in comparison to other virus groups it
could be related to the presence of defective interfering
(DI) particles. It has been shown for VSV that DI particles not only interfere with the replication of infectious
virus in tissue culture, but are also capable of modifying
an acute infection to a more subacute disease process (41).

On the other hand, the recorded biochemical and immunological differences between measles and SSPE virus suggest
that SSPE virus may possess additional properties which
favour persistency. The observed differences between the
M proteins of these viruses indicate one possible site at
the replication cycle, where interference can occur. If
SSPE virus has been derived from measles virus by mutation
it is conceivable that the gene region coding for the M
protein is affected and a non-functioning M protein could
be produced which may block virus assembly. This could
result in the initiation and maintainance of a non-productive persistent infection in which no virus particles
are formed. This hypothesis is compatible with the accumu-
lation of nucleocapsids found in infected brain cells and
the failure to detect virus budding.

Future virological investigations will have to analyse in
detail the virus-host relationship in SSPE. It will be
necessary to further characterize the virus RNA from SSPE
brain material to obtain direct information about the
state of viral defectiveness in brain cells. Moreover,
one should attempt to define the virus specific proteins
present in brain cells and if possible to identify those
in the cell membranes. A detailed analysis of the immune
response to measles virus structural proteins in SSPE
patients could provide valuable information about the synthesis of these proteins and indirectly their role in the
disease process. In addition, studies in measles infected
tissue cultures have shown that extra cellular measles
virus antibodies block membrane-incorporated viral proteins and thus prevent the immune surveillance system
from eliminating these infected cells (42). This event
has been postulated as one possible mechanism by which

measles virus persistency in the central nervous system is
maintained. Moreover, measles antibodies may also interfere by membrane signals with intracellular virus assembly
which could result in a state of virus defectiveness. An
investigation of these virological questions may lead to a
better understanding of the pathogenetic mechanism responsible for this chronic CNS infection.

Acknowledgements

The experimental work presented is supported by Deut-
sche Forschungsgemeinschaft. We wish to thank Magdalene
Pföhler for excellent technical assistance and Helga
Schneider for typing the manuscript.

REFERENCES
1. Meulen, V ter, M Katz, D Müller: Subacute sclerosing panencephalitis. In: Current Topics in Microbiology and
   Immunology, Arber et al (eds), Berlin, Heidelberg, New
   York: Springer Verlag, 1972, 57, p1-38.
2. Agnarsdottir, G: Subacute sclerosing panencephalitis.
   In: Recent Advances in Clinical Virology, AP Waterson
   (ed) Edinburgh, London, New York: Churchill Livingstone, 1977, p21-49.
3. Dawson, JR: Cellular inclusions in cerebral lesions of
   lethargic encephalitis. American Journal of Pathology
   9: 7-16, 1933.
4. Bouteille, M, C Fontaine, C Vedrenne, J Delarue: Sur un
   cas d'encéphalite subaigue à inclusions. Étude anato-
   moclinique et ultrastructurale. Revue Neurologique
   113: 454-458, 1965.
5. Katz, M, H Koprowski: The significance of failure to isolate infectious viruses in cases of subacute sclerosing panencephalitis. Archiv für die gesamte Virusforschung 41: 390-393, 1973.
6. Horta-Barbosa, L, R Hamilton, B Wittig, DA Fuccillo,
   JL Sever, ML Vernon: Subacute sclerosing panencephalitis: isolation of suppressed measles virus from lymph
   node biopsies. Science 173: 840-841, 1971.
7. Doi, Y, T Samse, M Nakajima, S Okawa, T Katoh, H Itoh,
   T Sato, K Oguchi, T Kumanishi, T Tsubaki: Properties
   of a cytopathic agent isolated from a patient with
   subacute sclerosing panencephalitis in Japan. Japanese
   Journal of Medical Science and Biology 25: 321-333,
   1972.
8. Kratzsch, V, WW Hall, K Nagashima, V ter Meulen: Biological and biochemical characterization of a latent subacute sclerosing panencephalitis virus infection in
   tissue cultures. Journal of Medical Virology 1: 139-
   154, 1977.

9. Mehta, PD, A Kane, H Thormar: Quantitation of measles virus-specific immunoglobulins in serum, CSF and brain extract from patients with subacute sclerosing panencephalitis. Journal of Immunology 118: 2254-2261,1977.
10. Vandvik, B, E Norrby: Oligoclonal IgG antibody response in the central nervous system to different measles virus antigens in subacute sclerosing panencephalitis. Proceedings of the National Academy of Sciences of the United States of America 70: 1060-1063, 1973.
11. Kreth, HW, H Pabst: Recent Findings on Cell-Mediated Immune Reactions in Acute Measles and SSPE (this publication).
12. Kingsbury, DW, MA Bratt, PW Choppin, RP Hanson, Y Hosaka, V ter Meulen, E Norrby, W Plowright, R Rott, WH Wunner: Paramyxoviridae. Intervirology 10: 137-152, 1978.
13. Waterson, AP: Measles virus. Archiv für die gesamte Virusforschung 16: 57-80, 1965.
14. Norrby, E: Separation of Measles virus components. Archiv für die gesamte Virusforschung 14: 306-318, 1964
15. Underwood, B, F Brown: Physico-chemical characterization of Rinderpest Virus. Medical Microbiology and Immunology 160: 125-132, 1974.
16. Oyanagi, S, V ter Meulen, M Katz, H Koprowski: Comparison of subacute sclerosing panencephalitis and measles viruses: an electron microscope study. Journal of Virology 7: 176-187, 1971.
17. Schluederberg, A: Measles Virus RNA. Biochem biophys Res Commun 42: 1012-1015, 1971
18. Mountcastle, WE, PW Choppin: A Comparison of the Polypeptides of Four Measles Virus Strains. Virology 78: 463-474, 1977.
19. Graves, MC, SM Silver, PW Choppin: Measles virus polypeptide synthesis in Infected Cells. Virology 86: 254-263, 1978.
20. Tyrrell, DLJ, E Norrby: Structural Polypeptides of Measles Virus. J gen Virol 39: 219-230, 1978.
21. Hall, WW, WR Kiessling, V ter Meulen: Biochemical comparison of measles and subacute sclerosing panencephalitis viruses. In: Negative Strand Viruses and the Host Cell, RD Barry, BWJ Mahy (eds), London, New York: Academic Press, 1977,p144-156.
22. Lamb, RW, PW Choppin: The Synthesis of Sendai virus polypeptides in Infected cells. Virology 81:382-397,1977.
23. Yeh, J, Y Iwasaki: Isolation and characterisaion of subacute sclerosing panencephalitis nucleocapsids. Journal of Virology 10: 1220-1227, 1972.
24. Hall, WW, V ter Meulen: RNA homology between subacute sclerosing panencephalitis and measles virus. Nature London 246: 474-477, 1976.
25. Hall, WW, WR Kiessling, V ter Meulen: Studies on the membrane protein of subacute sclerosing panencephalitis and measles viruses. Nature London 272:460-462,1977.
26. Wechsler, SL, BN Fields: Differences between the intracellular polypeptides of measles and subacute sclerosing panencephalitis virus. Nature,London 272:458-460, 1978.

27. Rima, BK, SJ Martin, EA Gould: A Comparison of Polypeptides in Measles and SSPE Virus Stains. J gen Virol 42: 603-608,1979.
28. Stephenson, JR, V ter Meulen: The Virological State in Subacute Sclerosing Panencephalitis. NATO Advanced Study Institute on Humoral Immunity in Neurological Diseases, Plenum Press, 1978.
29. Adams, JM: Comparative study of Canine Distemper and Respiratory disease of Man. Pediatrics 11: 15-27, 1953.
30. Imagawa, DT: Relationships among measles, canine distemper and rinderpest viruses. Progress in Medical Virology 10: 160-193, 1968.
31. Sato, TA, K Yamanouchi, A Shishido: Presence of Neutralizing Antibody to Canine Distemper Virus in Sera of Patients with Subacute Sclerosing Panencephalitis. Archiv für die gesamte Virusforschung 42: 36-41, 1973.
32. Yamanouchi, K, F Kobune, A Fukuda, M Hayami, A Shishido: Comparative Immunofluorescent Studies on Measles, Canine Distemper, and Rinderpest Viruses. Archiv für die gesamte Virusforschung 29: 90-100, 1970.
33. Örvell, C, E Norrby: Further Studies on the Immunologic Relationships among Measles, Distemper, and Rinderpest Viruses. The Journal of Immunology 113: 1850-1858, 1974.
34. Horta-Barbosa, L, DA Fuccillo, R Hamilton, R Traub, A Ley, JL Sever: Some characteristics of SSPE measles virus. Proceedings of the Society for Experimental and Biological Medicine 134: 17-21, 1970.
35. Meulen, V ter, M Katz, YM Käckell, G Barbanti-Brodano, H Koprowski, EH Lennette: Subacute Sclerosing Panencephalitis: In Vitro Characterization of Viruses Isolated from Brain Cells in Culture. The Journal of Infectious Diseases 126: 11-17, 1972.
36. Ecob, MS: PhD thesis, University of Cambridge, 1975.
37. Hamilton, R, L Barbosa, M Dubois: Subacute sclerosing panencephalitis measles virus: study of biological markers. Journal of Virology 12: 632-642, 1973.
38. Meulen, V ter, WW Hall, HW Kreth: Pathogenic aspects of subacute sclerosing panencephalitis. In: Persistent Viruses, CJ Stevens, GJ Todaro, CF Fox (eds), New York: Academic Press, p615-633, 1978.
39. Gatti, JM, RA Good: The immunological deficiency diseases. Medical Clinics of North America 54: 281-307, 1970.
40. Burnet, FM: Measles as an index of immunological function. Lancet ii: 610-613, 1968.
41. Huang, AS: Viral pathogenesis and molecular biology. Bacteriological Reviews 41: 811-821, 1977.
42. Joseph, BS, MBA Oldstone: Immunologic injury in measles virus infection. II. Suppression of immune injury through antigenic modulation. Journal of Experimental Medicine 142: 684-876, 1975.

MEASLES VIRUS-HOST CELL RELATIONSHIPS
IN SUBACUTE SCLEROSING PANENCEPHALITIS

K. B. FRASER

SUMMARY

In SSPE measles virus is present as a persistent infection showing
varying degrees of incomplete maturation. By manipulation measles
virus can be made to display all the biological behaviour of SSPE
virus including rapid adaptation to the cell – associated state. The
possible origins of the persistence of measles virus are
physiological, genetic, infectious or immunological. Animal
experiments point to a failure of cell – mediated immunity in the
presence of adequate humoral immunity as the most favourable
circumstances for chronic infection to begin. This state of affairs
could arise from immaturity or from immunosuppression by virus
infection.

INTRODUCTION

Subacute sclerosing panencephalitis (SSPE) is a rare and late sequel
of measles. By the time that the illness has begun, the numerous
infected cells in the brain have overwhelmed or bypassed the immune
clearing system much in the same way as a palpable tumour has got
beyond the control of a defence which would normally be capable of
disposing of a few discrete malignant cells. The immunological
relationships are probably equally complex in the late stages of the
two illnesses, but the specific immune reactions against measles
virus are said to be largely intact at the beginning of the subacute
sclerosing panencephalitis (1).

Distemper virus has been suspected on epidemiological grounds
(2, 3) as a cause of SSPE alternative to or additional to measles
virus but an early and important contribution by Connolly (4)
(Table 1) rules out canine distemper as the characteristic antigen in
brain during SSPE. The serological tracers demonstrate clearly that,
if distemper virus plays any part in the aetiology of SSPE, either
independently or in combination with measles virus, at least it is
not expressed as fluorescent – staining antigen in the central
nervous system.

Table 1.   IMMUNOFLUORESCENCE OF SSPE BRAIN,
           DISTEMPER BRAIN AND VIRUS - INFECTED CELLS

	FITC - Measles Antisera		FITC - Distemper Antiserum
	Patient	Goat	Horse
SSPE Brain	++	++	−
Healthy Control Brain	−	−	−
Distemper Brain	++	+	++
Control Dog Brain	−	−	−
Measles Infected Cells	++	++	−
Uninfected Cells	−	−	−
Distemper Infected Cells	++	+	++
Uninfected Cells	−	−	−

+ etc. = Intensity of fluorescence. Adapted from Connolly, 1968.

Other viruses have been suspected of being associated with SSPE.
Electron microscopy has occasionally demonstrated papova virus - like
particles in SSPE brain (5) and herpes - virus - like particles in
measles encephalitis (6) giving rise to the suggestion that SSPE is
dependent on double infection, but so far only antibody to Epstein
Barr virus of the herpes virus group has been reported in addition to
measles virus antibody as being significantly associated with SSPE.

Table 2.   NUMBER OF SSPE PATIENTS WITH EBV ANTIBODY

Present In		Ref.
SSPE	Non - SSPE	
8/8	36/88	43
35/45[1]	39/54	25
19/41[2]	0/20	
16/16[1]	123/140	*
15/16[3]	N.D.	

1 = VCA; 2 = EA; 3 = EBNA - specific antibody; N.D. = not done;
* = Haire & Connolly unpublished.

All patients out of 8 tested had it (7) on one survey and titres are
said to be raised above average in another (8) but as antibody to
this virus is raised during measles (9) and as both viruses can
inhabit lymphocytes, there may be some other explanation for the
association than double infection of brain. Our figures in Belfast,
showing 15/16 patients with more than one antibody to EB virus

components support these findings (Table 2) (Haire and Connolly, unpublished).   Many more data are needed to substantiate these claims and to assess their significance.   It seems unlikely that simultaneous double infection by measles virus and another different virus is the explanation for chronic encephalitis.

A PERSISTENT AND DEFECTIVE INFECTION

Persistence - Measles virus or variants of it have now been recovered frequently from SSPE brain.   When the presence of measles virus in SSPE became known, the interest of virologists was directed to Rustigian's much neglected contribution to virology.   He had shown that measles virus could readily enter into a perpetuating, non - cytocidal infection in human cell culture.   Adding measles antibody to the medium not only helped to stabilise the infected culture, but also helped to establish a replicative cycle which made antigen but produced no or very little, infectious virus (10, 11).   At that time few virologists were in a position to think of incomplete growth cycles in terms of a limited number of virus structural proteins.   In Rustigian's system transmission of infection was vertical at cell division without evidence of spread by cell - fusion.   This could not apply to infected neurones, which do not divide.

The next important improvement in our understanding of the pathogenesis of SSPE was the discovery that free measles virus was rarely present *in vivo* , but that culture of live brain cells or co - cultivation with a species of cell susceptible to measles virus regularly showed cell - associated measles virus antigen with or without giant - cell formation and from these cultures free measles virus could sometimes be recovered (12, 13, 14).

Table 3.

SSPE VIRUSES
VARIATION IN STAGE OF RECOVERY OF HAEMAGGLUTININ

Patient	Pass	HA	Ref.
1	1 and 2	+	(12)
2	7	+	(12)
3	16	+	(13)
4	200	0	(18)[*]
5	55	0	(19)
6	Repeated co-cultivation	0	(20)

+ = recovered; 0 = not recovered at any time; * = quoted by (20).

It was apparent from the beginning that the replication of measles virus was under some form of physiological control from which it took different lengths of time or different numbers of passage in tissue culture before the virus was released to give anything like normal laboratory behaviour (Table 3). This recovery took place long after any immunological inhibition present in the starting material would have been diluted out in culture. The properties of stable SSPE viruses varied as much as do those of laboratory strains of standard measles virus and the existence of a specific SSPE type of measles virus seemed unlikely (15, 16). Later, cultures were recovered which alternated between lytic and non - lytic cycles of infection (17) and several have also been reported as being permanently non - virus - yielding (Table 3) but nevertheless function as cytopathogenic infectious centres and cause virulent encephalitis *in vivo* (18,19,20).

A Defect of Maturation - The nature of the non - productive state has been best revealed by electron microscopy. It is a blocking of maturation at various stages after the synthesis of virus nucleocapsids even in the presence of structural proteins of the virus. There is one reported instance of a culture in which all but 2 structural polypeptides of the measles virus are absent (21) and at the other extreme another showing apparently complete, but non - infectious, virions (22). Normally at the peak of the normal virus growth cycle rough nucleocapsid tubules of measles virus are abundant and generally distributed in the cytoplasm with a noticeable tendency to be aligned at the inner surface of the plasma membrane where spikes of haemagglutinin are beginning to appear. Budding virions are numerous at or near the cell surface. In SSPE and in persistent defective cultures there is little budding and particles are spikeless or devoid of a proper complement of nucleo - capsid. In the cytoplasm nucleocapsid molecules often remain aggregated and are not associated with the sub - membranal surface (23, 24, 25, 26). Reorganisation of nucleocapsid and migration to the sub - membranal area may begin as soon as permissive cells fuse with carrier cells (27). These morphological points are very similar to stable steady - state infections where concentration of antigen in aggregates in the cytoplasm and diminution of virus antigens at the cell surface are characteristic (28, 29). The growth of measles virus in the carrier state is often temperature - sensitive (30, 31, 32).

Following these observations and considerations the virologist is faced with five main questions: (1) Why is the disease rare? (2) What happens initially to start off the persistent virus - cell relationship? (3) What makes the cycle defective? (4) How does defective virus spread from cell to cell in the presence of antibody? (5) Why does cell-mediated immunity not eliminate infected nerve cells?   The answers will always depend on reasonable speculation, proof being impossible, but during the last few years much has been discovered about measles virus and its immunology which makes the various possibilities actual rather than theoretical.

GENESIS OF PERSISTENT INFECTION

Epidemiology - Patients who get SSPE have no stigmata of immune deficiency before the onset of illness such as has been associated with giant cell pneumonia (33) so we must conclude that chance or some inate unrealised condition or some unobserved environmental event determines the difference between normal recovery from childhood measles and the origin of a chronic submerged process which surfaces at an average of five years (2 to 20 years) later.   Post - vaccine SSPE has a somewhat shorter incubation period, about 3 years (34).   In view of the solid immunity that follows measles, SSPE is not likely to be a second measles infection unless strains of SSPE virus are proven to be serologically different from measles.   Neither epidemic clustering nor studies of SSPE in twins support the idea of a specially neurotropic strain of SSPE measles virus nor of an inborn defect of resistance to measles virus but point rather to some environmental influence (35).

The one significant piece of epidemiological evidence about which all observers are agreed is that half the patients who get SSPE had measles below the age of 2 years, well below the modal age of infection for their own social group (34, 36).   There is a pre-dominance of male patients, but the reason for it is not known.

Physiological Chance - The chance of SSPE beginning may be entirely dependent on a virus - cell relationship and may be physiological or genetic.   Persistent infection may be as natural a mode of existence for measles virus as lytic or cytocidal infection.   Given alternative pathways of replication in identical cells and in constant environmental conditions, one would expect a virus to have a definite probability, as does lysogenic bacteriophage, of taking one way or the

other. This probability will vary with the strain of host cell.
There are no measurements of such a probability with measles virus,
but alternative pathways of lysis or persistence certainly exist.

Table 4. GROWTH OF MEASLES VIRUS IN BRAIN CELL CULTURE

Species	Cell Culture	Measles Virus	SSPE Virus	Type of Growth	Ref.
Man	Astrocyte	+	N.D.	Persistent induced	(37)
Man	Astrocytoma	+	N.D.	Persistent induced	
Rat	Schwannoma	+	+	Persistent[1] immediate (ts)	(38)
Rat	Astrocytoma	+	+	Persistent[1] immediate (ts)	

Induced = by selection of survivors of lysed culture;
ts = temperature sensitive replication; 1 = host - controlled.

Brain cells are known to support virus growth readily without lethal
results and measles virus is no exception (Table 4). Different lines
of brain cells respond differently (37, 38) and persistent infection
may be host - controlled and temperature - sensitive (38). Since
there is indirect evidence from electroencephalograms that measles
virus probably invades the brain of half the children who get measles
(39) the opportunities for persistent infection to arise must be many.

Genetic Chance - The other sort of finite chance is mutation and it
will apply only if mutants of measles virus exist which normally cause
persistent infection. Proof requires plaque purification and that
rules out the investigation of defective mutants. Accordingly, some
workers have sought for and recovered variants from measles and SSPE
viruses which grow poorly *in vitro* and which make plaques of widely
different appearance (40) (Table 5). No constant link with SSPE has
been demonstrated but it is possible that strains showing a
combination of two markers, poor replication and good syncytial
formation tend to induce chronic encephalitis readily. Thormar makes
an excellent case for syncytial formation in cultured brain cells and
non - productive replication being linked to encephalitogenic potency
(Table 6) but he himself points out that encephalitic producer strains
may be made apparently non - neurotropic as a consequence of rapidly
inducing local concentrations of antibody in the central nervous
system (41).

Table 5.   PLAQUE VARIANTS OF MEASLES VIRUS

Plaque Type	Yield of Virus pfu	Encephalitis in Hamster		Ref.
Measles				
Small	$10^{5.0}$	13/59	(22)	⎤
Large	$5 \times 10^{3.0}$	33/69	(47)	(40)
Small	$7 \times 10^{3.0}$	0/38	(0)	⎦
SSPE-1				
Small	$5 \times 10^{7.0*}$	12/12	(100)	⎤
Large	$1 \times 10^{4.0}$	1/10	(10)	(+)
Points focus	$5 \times 10^{5.0}$	0/10	(0)	⎦

* = assayed by fluorescent antibody; 1 pfu = 1 focus forming unit; + = Gould et al unpublished.

Table 6.   SYNCYTIAL FORMATION IN VITRO AND ENCEPHALITIS

Measles-Virus		HAD	Virions (EM)	Syncytia in Vitro (FB)	Encephalitis in Ferret
Wild types	1	+	+	−	−
	2	+	+	−	−
SSPE	1	+	+	−	−
	2	+	+	−	−
	3	+	+	−	−
SSPE	4	+	−	+	+
	5	±	−	+	+
	6	−	−	+	+
	7	−	−	++	++

HAD = haemadsorption; FB = Ferret brain culture (modified from 41).

One unusual feature of two of Gould's variants recovered from a well known strain of SSPE virus (SSPE 1. Horta - Barbosa) is their restricted growth in the absence of methionine which they do not incorporate adequately into virus protein (Fig. 1). How this may be related to defective growth or to intra - cerebral habitat has not yet been investigated, but it is of consequence that the small focus - former is almost undetectable without fluorescent antibody assay and may have been missed in many recorded, non - producing SSPE strains. The same SSPE virus also yielded a productive variant which did incorporate methionine as well as laboratory strains of wild type

virus did.   Thus, non - cytocidal variants of measles do exist, can
cause encephalitis and, if present, would be expected, even more than
wild type virus, to set up persistent infection in the human brain.

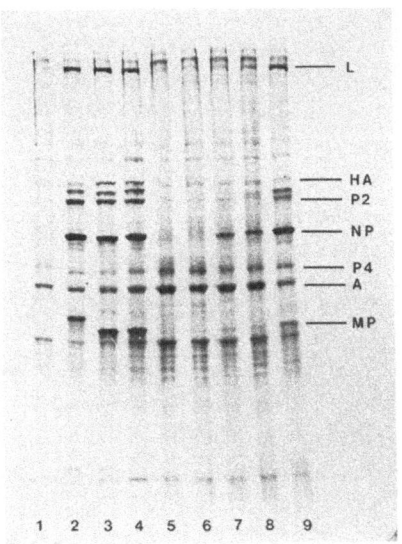

Fig. 1.   SDS - PAGE analysis of ^{35}S methionine labelled
measles and SSPE virus variants.   1. Uninfected Vero cells.
2. Human measles isolate.   3. Edmonston virus.   4, 5 and 6.
Small plaque, Foci - producing and large plaque variants,
respectively, of SSPE - 1 virus isolated by Horta - Barbosa.
7, 8 and 9. The dilute large plaque, undilute large plaque
and small plaque variants of Edmonston virus described by
Chiarini et al (40).   HA - haemagglutinin, NP - nucleoprotein,
A - actin and MP - membrane protein.

Metabolic Abnormality - The inhibition of maturation of SSPE viruses
could be produced by a control of cell metabolism, for the production
of infectious measles is recognised to require some unknown function
of the cell nucleus (42).   Cell fusion is very active in early
isolates of SSPE viruses so fusion factor, which is probably virus

haemolysin, must be an important constituent or product of SSPE
defective virus strains.   It is interesting therefore that when
Shirodaria (43) preferentially inhibited the synthesis of measles
virus haemolysin by using low concentrations of glucose analogues,
either glucosamine or 2 deoxy D glucose (Fig. 2) he and Dr. Dermott
observed that maturation of virus ceased and nucleocapsids remained
aggregated in the cytoplasm instead of lying adjacent to the
differentiating cell membrane.   They have also shown that removal of
the chemical inhibitor was followed by disaggregation of nucleocapsid
clusters and rapid resumption of maturation within 2 hr. (Fig. 3).   It
cannot be decided yet whether maturation is prevented by lack of
haemolysin or whether migration of components in the cell and protein
synthesis were both indirectly inhibited by loss of some other
metabolic pathway, but the general picture is very like that in
replication by defective SSPE measles virus.

Multiple Infections (DI Particles) - It has been shown that most
persistent virus infections are associated with large numbers of
particles that contain incomplete amounts of RNA and do much to help
the establishment of non - cytocidal infection (44, 45). With measles
also there is good evidence that a susceptible cell - culture can be
infected non - cytocidally by measles virus prepared by undiluted
passage (46) and so carrying a high proportion of these DI particles
(47). Even after 93 passes, Rima's culture contained no defective or
temperature-sensitive virus, which indicates that noncytocidal (46)
infection is not dependent on the appearance of poorly - growing or
ts variants like those recovered from SSPE virus (Table 5).   If this
mechanism of interference is responsible for initiating persistent
infection in SSPE, some system of multiple infection of brain cells
must be postulated.   Measles spreads in the patient in mononuclear
leukocytes in which it can replicate (48, 49) so that inter - cellular
transmission from leukocyte to brain cell is likely to bring in a high
multiplicity of genomes.   Alternatively, as suggested by Martin (16)
the measles virus itself could be polyploid and so produce the same
polygenic infection as a virus - loaded white blood cell.

Antigenic Modulation - Persistent infection and defective growth of
the virus are both accelerated by the presence of measles virus -
specific antibody in the culture medium (10, 11).   Recently the
consequences of antibody uniting with virus antigen at the cell surface

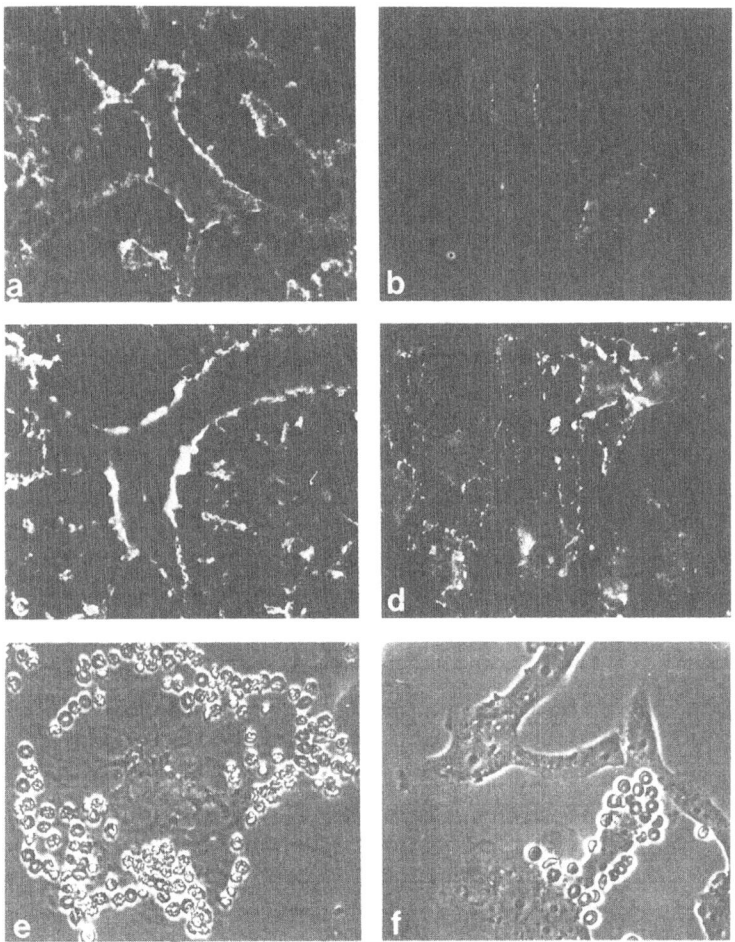

Fig. 2. Immunofluorescence of and haemadsorption by measles
virus antigens in unfixed Vero cells infected with strain TC243,
incubated without (Left) and with (Right) 4mM̄ 2-deoxy-D-glucose
added at the beginning of the incubation period.
a, b - haemolysin; c, d - haemagglutinin; e, f - haemadsorption.
Note the loss of antigen, especially haemolysin, loss of
syncytial formation and reduction of haemadsorption in presence
of 2-deoxy-D-glucose.

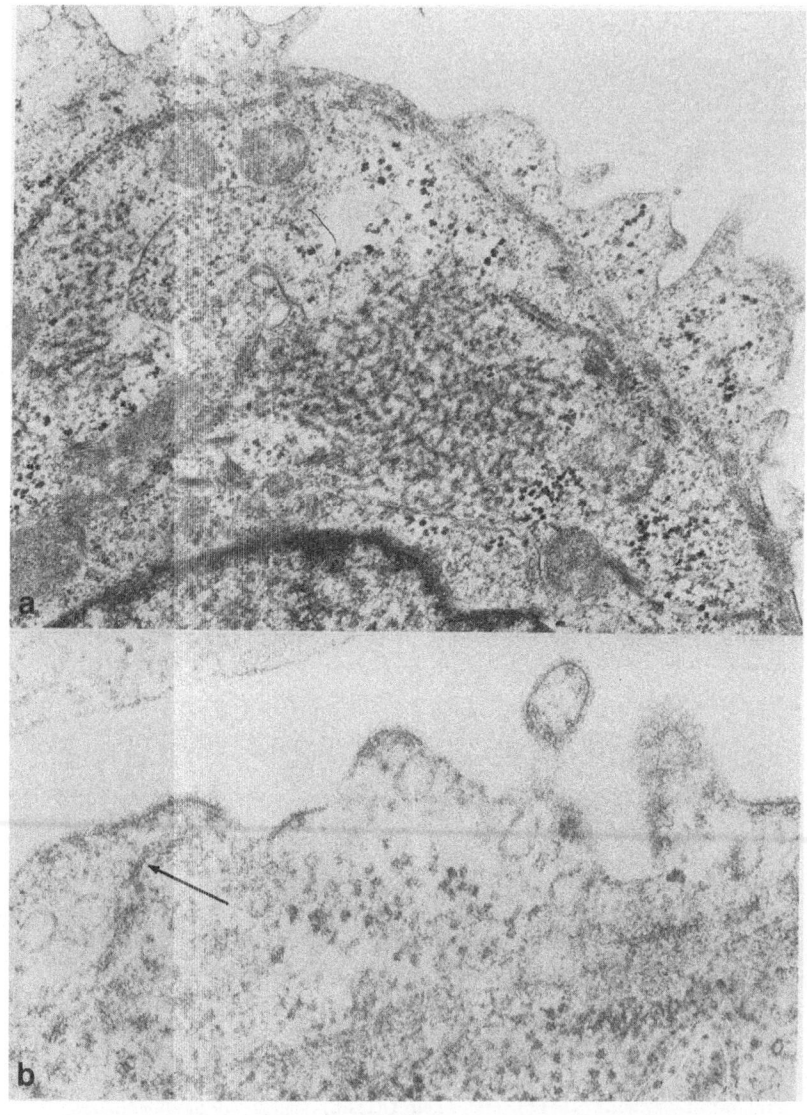

Fig. 3. (a) HEp2 cells in 10mM̄ 2-DG 26 hr. after infection with measles virus. Cytoplasmic inclusion of rough nucleocapsids (NC); no submembranal alignment; no budding, X 40,000; (b) without 2-DG, NC migrates towards cell surface (arrow) where differentiation and budding are seen, X 90,000.

have been described, both as capping, which means moving to a
restricted site on the surface of the cell (50) and stripping which
leads to loss of antigen from the cell surface.   Such antigenic loss
or modulation can provide an explanation for incomplete maturation
because the surface antigens required by the virus at budding are
blocked by antibody (51) or are continually removed and unused
nucleocapsid will build up within the cell.   The process has also
been credited with protection of infected cells which, being stripped
of virus antigen, are not subject to attack by measles virus –
specific immune processes (52).   It should be remembered, however,
that reduction of surface antigen ensues in many persistent infections
without the intervention of antibody.

  So, in addition to inherited virus traits, there are two other
mechanisms which result physiologically in the establishment of a
persistent infection by measles virus, either multiple infection with
D.I. particles or antigen modulation.   The latter method does not
explain how defectiveness persists after antiserum has been removed
from the system; the former method is self – perpetuating because D.I.
particles are replicated by each infected cell.   It is difficult to
explain how either method should select SSPE virus which retains its
ability to spread and to cause cell death *in vivo*.   The formation of
syncytia by SSPE carrier cells *in vitro* is inhibited by measles
antiserum.   Anatomy should not be neglected.   It may be that SSPE
virus is selected in situ by its ability to spread entirely through
synapses as suggested by patterns of fluorescent antibody staining
(53, 21) and the immune system does deal capably with all virus
antigens at exposed sites.

ANIMAL EXPERIMENTS AND IMMUNOLOGICAL CONTROL

Animal experiments provide information about multi – cellular
relationships.   They have been used to show that SSPE viruses can
cause chronic encephalitis which is histologically similar to SSPE;
they have shown that standard strains of measles virus can in certain
circumstances initiate chronic encephalitis remarkably similar to SSPE
(Table 7).   Most important of all, they have revealed that the onset
and course of measles virus – induced encephalitis are firmly
controlled by the immune state of the host. (Table 8).

  Extra cellular SSPE virus is hardly ever neuro – virulent for
adult animals after intra – cerebral inoculation (54).   By far the
most reliable method of inducing encephalitis in all species of

laboratory animal is to inoculate cell - associated virus in living cells (55, 56, 57). This finding is in accord with the cell - fusing properties of isolated SSPE strains and fusion has actually been demonstrated *in vivo* both in animals and man (58, 59). Measles specific - antibody is not always formed after such inoculation even when encephalitis has followed (60) which indicates that measles - specific antibody is not the mediator for the production of chronic encephalitis nor for its symptoms. Nevertheless the time - course of infection does depend on the virus - immune state as seen, for example, by inoculating neuro - adapted SSPE virus into rhesus monkeys in whom immunity was suppressed by cyclophosphamide (61).

Table 7.

ENCEPHALITIS IN HAMSTERS
SSPE AND MEASLES VIRUSES

Virus	Type of Illness	Inclusion Seen	Ref.
SSPE	Acute	+	(75)
SSPE	Acute	+	(76)
SSPE	Chronic	+	(64)
Vaccine Measles	Chronic	−	(77)
Measles Carrier Cells	Chronic	+	(78)

Table 8.

EFFECT OF IMMUNE STATE ON EXPERIMENTAL
ENCEPHALITIS INDUCED BY MEASLES VIRUS

Immunosuppressed	Virus	Host	Result	Ref.
Cyclophosphamide	Measles Vaccine HNT	Monkey	Susceptible[1] to HNT only	(61)
Cyclophosphamide	Measles Vaccine	Hamster	Reactivation	(77)
Anti-Lymphocyte	SSPE − HBS	Hamster	Susceptible[1]	(62)
Anti-Thymocyte Serum	SSPE − LEC	Hamster	Susceptible[1]	(66)
Neonatal Thymectomy	SSPE − HBS	Hamster	Susceptible[1]	(65)
Hypersensitized (Allergic)	Measles Neurotropic	Rat	Protective[2]	(79)

1 = compared with non - suppressed controls.
2 = compared with non - sensitized controls.

Partial immunosuppression by anti - lymphocyte serum can convert a
non - lethal infection with SSPE virus into a lethal one (62) and the
morphogenesis of the virus in chronic infection of weanling hamsters
is defective as compared with complete maturation in the acute
infection of newborn animals (63). Age of the host has an influence
on the chronicity of infection by SSPE virus, as it has with
laboratory strains of measles virus. Infection was chronic and
focal with intranuclear inclusion bodies in twelve week - old
hamsters, but acute and inflamed in three week - old animals (64).
The same strain of virus was more virulent in adult hamsters which had
had a neonatal thymectomy (65) or which been treated with anti -
thymocyte serum (66) than it was in untreated hamsters. Further work
on non - immune rhesus monkeys has shown that inoculated extra -
cellular SSPE viruses was harmless, but 3 of 4 non - productive SSPE
viruses produced severe acute encephalitis. Slightly neuro -
adapted non - productive SSPE virus produced a chronic encephalitis
in animals that were already immunized against measles (67).

Speed of adaptation to defective growth has been formally
demonstrated. A fully lytic, productive and neurotropic derivative
of SSPE virus could be shown to produce encephalitis in weanling
hamsters. The parent virus could be recovered from inoculated
brains up to about the eighth day after inoculation when measles virus
- specific antibody began to appear. Thereafter the only means of
demonstrating virus infectivity was by co - cultivation of brain
cells. Although HA and HL were both inserted into the cell membrane,
no infectious virus could be released by freezing and thawing.
Adaptation to defectiveness had occurred as early as 8 days after
inoculation (68).

Intra - cerebral inoculation is a highly artificial procedure but
encephalitis has been shown to follow parenteral inoculation -
intraperitoneal in the hamster (69) and intramuscular, using a non -
producer strain, in the monkey (70). The last experiments justify
our speculation about transmission of infection at high multiplicity
by means of leukocytes (p.) since the SSPE virus used could not
travel extra - cellularly from muscle to brain.

The slow progress of SSPE may have other causes than immunological
constraint. Encephalitis due to laboratory strains of measles virus
is greatly reduced in virulence as age of the host increases (71, 53,
72). A non-SSPE, hamster - neurotropic strain of measles virus

which is rapidly restricted in its effects on mice after they are 6 to
7 days old has been found to make much virus antigen at the
restrictive age although no virus components are formed; no extension
beyond the primarily infected cells is possible (73). The restriction
is not due to immune reactivity, to interferon production, to body
temperature or to the influence of cellular proteases on the
infection (74).   If such processes occur in human SSPE the slow
progress of virus throughout the brain would be partly explained.   A
small population of producer cells would gradually fill the blind
alleys with non - infectious but lethal, virus antigen.

RECAPITULATION AND ARGUMENT

Subacute sclerosing encephalitis, SSPE, is assumed here to be a sequel
of measles and not a reinfection, in conformity with epidemiological
observations, but the virus - host relationships discussed could
equally well arise as the result of a reinfection that had by - passed
normal host defences.   There is no evidence for this.

The nature of the encephalitis and the properties of persistent
and defective infection with measles virus indicate five main problems
in the pathogenesis of the disease.   They are the epidemiology, the
origin of persistent infection, the development of defective virus
reproduction, the means of spread of non - productive but lethal
measles virus in brain, and failure of elimination by the immune
system.

There is reasonable evidence that the encephalitis is essentially
an infection by a single virus, measles virus.   There is no
consistent evidence that SSPE is a continuing double intracerebral
infection with measles virus and another such as distemper virus,
papova viruses or herpes viruses.   Coincidental but temporary
infection in the brain or elsewhere cannot be ruled out as part of the
aetiology.

Persistent infections by wild - type measles virus often become
defective, especially with the aid of measles virus - specific
antibody in the culture medium and the properties of carrier cultures
and of virus recovered from them strongly resemble cell - associated
and intracellular strains of SSPE virus.   In both experimental and in
SSPE systems defectiveness is revealed to be a failure of maturation
of various degrees of incompleteness but often relieved by fusion with
permissive cells.

Persistence of measles virus can arise readily and spontaneously especially in many kinds of brain cells where it is host – controlled at low multiplicity or is rather similar to artificial persistent virus cell systems induced at high multiplicity.   Both kinds may show temperature – sensitive synthesis of virus.

Genetic variants of measles virus exist which have a propensity to set up non – cytocidal rather than lytic infection of cells in culture.   They tend to be missed in tests for infectious virus and could act like D.I. particles in suppressing maturation of lytic virus.   It is also possible to stimulate the physiology and morphology of defective infection by inhibition of normal growth cycle biochemically.   There is no known counterpart in SSPE.

Antigenic modulation or stripping of measles antigen from the surface of virus – infected cells is common to experimental persistent infections and perhaps to *in vivo* 'protection' of the infected host – cell from cell – mediated immune clearance.   The immune modulation mechanism does not explain why spread by fusion is not prevented *in vivo* when, in fact, antibody *in vitro* inhibits syncytial formation quite well.

Animal experiments have been invaluable in showing that encephalitic infections produced by wild type or laboratory measles viruses and by SSPE virus strains are remarkably alike.   Above all they have shown that the genesis and the course of chronic measles encephalitis can be determined by the immune state of the host.   This is seen in three sets of conditions: (1) Before specific active cell – mediated immunity exists but when protective antibody is present as in passive maternal immunity.   (2) After specific humoral and cell – mediated immunity is present, but cell – mediated immunity is suppressed.   (3) In one set of experiments only, where hyperergy to measles virus was established before infection.   The first two conditions determine increased susceptibility, the third increased resistance, to encephalitis.

The importance of other factors has been discovered by experiments in animals.   One is that adaptation to defectiveness occurs rapidly in the brain and is complete at the time when antibody begins to inhibit cell – to – cell spread of infection.   More important still, cell – associated, syncytial – forming virus is the most potent inducer of encephalitis and excites least immune response.

Why measles virus - specific antibody of high titre cannot inhibit
the spread of syncytial forming virus *in vivo*, when small amounts do
so *in vitro* seems paradoxical.   In all probability the facts simply
mean that virus spreads through cell synapses and the comparative
rarity of giant cells in SSPE means that antibody is effective
whenever the virus antigen is exposed at the cell surface.

A second finding is that increasing maturity affects susceptibility of cerebral tissue to measles virus and slowness of spread -
the slow virus infection - could be accounted for by the reproduction
of relatively minute amounts of complete infectious virus or by the
high probability of virus entering a series of blind alleys
represented by cells in which virus replication is abortive.

The most significant factor in the epidemiology of SSPE is the
early age at which measles had been acquired by at least half the
patients.   Two attributes of experimental measles encephalitis are
related to this.   Access to the brain could be easier in the very
young or immunological competence to deal with measles virus could be
less well developed.   Alternatively the leucocytes of very young
subjects could be more susceptible to measles virus or could produce
more virus than leucocytes of older children, so subjecting the nervous
system to a heavier invasion of transported virus.   It will never be
possible to prove whether one or more of these possible characteristics
of immaturity are operating at the time when the mechanisms of SSPE
are first set in motion, but they are less applicable to older cases
and they do not explain the entire epidemiology of SSPE.

One factor which is environmental, applicable to the age group
concerned as well as to older ages and immunologically acceptable,
is coincidental infection by a virus which can produce temporary
immunodepression of greater or lesser specificity for measles virus.
A 'two virus' hypothesis has been favoured amongst the problems of
virus - host relationships in multiple sclerosis (80) though without
virological or serological supporting evidence.   We have referred to
the serology in SSPE of Epstein - Barr virus (82, 83) which, amongst
others (81) can depress cell - mediated immunity.   In the older SSPE
patients who presumably have been immunologically mature, some reason
for immunodepression should be sought.   Coincidental virus infection
would be high on the list of such causes and further work on this line
of enquiry seems justified.

ACKNOWLEDGEMENTS

I am indebted to the following colleagues for discussion of their work and permission to use their data: J. Connolly, Table 1; M. Haire, Table 2; E. Gould and K. McCullough, Table 5; M. Gharpure and E. Gould, Fig. 1; P. Shirodaria, Fig. 2; and E. Dermott, Fig. 3. I thank Mr. R. Woods for photography.

REFERENCES

1. Agnarsdottir, G: Subacute sclerosing panencephalitis. In:
   Recent advances in clinical virology, Watson, A. (ed), London
   and Edinburgh, Churchill Livingstone, 1977, p21-49.

2. Brody, JA, R Detels: Subacute sclerosing pan-encephalitis: A
   zoonosis following aberrant measles. Lancet 2: 500-501, 1970.

3. Detels, R, JA Brody, J McNen, AH Edgar: Further epidemiological
   studies of subacute sclerosing panencephalitis. Lancet 2:
   11-14, 1973.

4. Connolly, JH: Additional data on measles virus antibody and
   antigen in subacute sclerosing panencephalitis. Neurology 18:
   87-89, 1968.

5. Koprowski, H, G Barbanti-Brodano, M Katz: Interaction between
   papova-like virus and paramyxovirus in human brain cells -
   a hypothesis. Nature 275: 1045-1047, 1970.

6. Dayan, AD: Encephalitis due to simultaneous infection by herpes
   simplex and measles viruses. J. neurol. Sci. 14: 315-323, 1971.

7. Joncas, J, G Geoffrey, B McLaughlin, G Albert, N Lapointe,
   P David, R Lafontaine, M Granger-Julien: Subacute sclerosing
   panencephalitis. Elevated Epstein-Barr virus antibody titres
   and failure of amantadine therapy. J. neurol. Sci. 21:
   381-390, 1974.

8. Gotlieb-Stematsky, GT, L Rannon, A Vonsovera: Antibodies to
   Epstein-Barr virus in subacute sclerosing panencephalitis
   patients. Europ. Neurol. 13: 418-421, 1975.

9. Gotlieb-Stematsky, T, L Rannon, A Vonsovera, N Varsano:
   Stimulation of antibodies to Epstein-Barr virus (EBV) in acute
   viral infections. Arch. Virol. 57: 199-204, 1978.

10. Rustigian, R: A carrier state in Hela cells with measles virus
    (Edmonston strain) apparently associated with non-infectious
    virus. Virology 16: 101-104, 1962.

11. Rustigian, R: Persistent infection of cells in culture by
    measles virus. II. Effect of measles antibody on persistently
    infected Hela sublines and recovery of a Hela clonal line
    persistently infected with incomplete virus. J. Bact. 92:
    1805-1811, 1966.

12. Horta-Barbosa, L, DA Fuccillo, JL Sever, W Zeman: Subacute
    sclerosing panencephalitis : Isolation of measles virus from
    a brain biopsy. Nature (London),221: 974, 1969.

13. Payne, FE, JV Baublis, HH Itabashi: Isolation of measles virus from cell cultures of brain from a patient with subacute sclerosing panencephalitis. New Eng. J. Med. 281: 585-589, 1969.

14. Degré, M, B Vandvik, T Hovig: Subacute sclerosing panencephalitis : Isolation and ultra-structural characterization of a measles-like virus from brain obtained at autopsy. Acta Path et. Microbiologica Scand. Section B. 80: 713-728, 1972.

15. Hamilton, R, L Barbosa, M Dubois: Subacute sclerosing panencephalitis measles virus : study of biological markers. J. virol. 12: 632-642, 1973.

16. Fraser, KB, SM Martin: Measles virus and its biology, London, Academic Press, 1977.

17. Burnstein, T, LB Jacobsen, N Zeman, T Tsu Chen: Persistent infection of BSC-1 cells by defective measles virus derived from subacute sclerosing panencephalitis. Infect. Immun. 10: 1378-1382, 1974.

18. Doi, Y, T Sampe, M Nakajima, S Okawa, T Katoh, H Itoh, T Sato, K Oguchi, T Kumanschi, T Tsubaki: Properties of a cytopathic agent isolated from a patient with subacute sclerosing panencephalitis in Japan. Jap. J. Med. Sci. Biol. 25: 321-333, 1972.

19. Makino, S, K Sasaki, M Nakagawa et al: Isolation and biological characterization of a measles virus-like agent from the brain of an autopsied case of subacute sclerosing panencephalitis. Jap. Microbiol. Immunol. 21: 193-205, 1977.

20. Mirchamsy, H, S Bahrami, A Shafyi, MS Shahrabady, M Kamaly, P Ahourai, J Razavi, P Nazari, T Derakshan, J Lotfi, K Abassioun: Isolation and characterization of a defective measles virus from brain biopsies of the patients in Iran with subacute sclerosis panencephalitis. Intervirology 9: 106-118, 1978.

21. Albrecht, P: Immune control in experimental subacute sclerosing panencephalitis. Amer. J. clin. Path. Suppl. to Vol. 70 (1): 175-184, 1978.

22. Kratsch, V, WW Hall, K Nagashima, V ter Meulen: Biological and biochemical characterization of a latent subacute sclerosing panencephalitis (SSPE) virus in tissue culture. J. med. virol. 1: 139-154, 1977.

23. Oyanagi, S, V ter Meulen, M Katz, H Koprowski: Comparison of subacute sclerosing panencephalitis and measles viruses: an electron microscope study. J. Virol. 7: 176-187, 1971.

24. Raine, CS, LA Feldman, RD Sheppard, LH Barbosa, MB Burnstein: Subacute sclerosing panencephalitis virus. Observations on a neuroadapted and non-neuroadapted strain in organotypic central nervous system cultures. Lab. Invest. 31: 42-53, 1974.

25. Dubois-Dalcq, M, K Worthington, O Gutenson, LH Barbosa: Immunoperoxidase labeling of subacute sclerosing panencephalitis virus in hamster acute encephalitis. Lab. Invest. 32: 518-526, 1975.

26. Dubois-Dalcq, M, TS Reese, M Murphy, D Fuccillo: Defective bud formation in human cells chronically infected with subacute sclerosing panencephalitis virus. J. Virol. 19: 579-593, 1976.

27. Knight, PR, RG Duff, R Glaser, F Rapp: Characteristics of the release of measles virus from latently infected cells after co-cultivation with BSC-1 cells. Intervirology 2: 287-298, 1973.

28. Walker, DL: The viral carrier state in animal cultures. Progr. med. Virol. 6: 111-148, 1964.

29. Fraser, KB: Defective and delayed myxovirus infections. Brit. Med. Bull. 23: 178-184, 1967.

30. Haspel, MV, PR Knight, RG Duff, F Rapp: Activation of a latent measles virus infection in hamster cells. J. Virol. 12: 690-695, 1973.

31. Gould, EA, PE Linton: The production of a temperature-sensitive persistent measles virus infection. J. gen. Virol. 28: 21-28, 1975.

32. Armen, RC, JF Evermann, AL Truant, LA Laughlin, JV Hallum: Temperature sensitive mutants of measles virus produced from persistently infected Hela cells. Arch. Virol. 53: 121-132, 1977.

33. Enders, JF, K McCarthy, A Mitus, WJ Cheatham: Isolation of measles virus at autopsy in cases of giant cell pneumonia without rash. New Engl. J. Med. 261: 875-881, 1959.

34. Modlin, JF JT Jabbour, JJ Witte, NA Halsey: Epidemiological studies of measles, measles vaccine, and subacute sclerosing panencephalitis. Pediatrics 59: 505-512, 1977.

35. Ch'ien, LT, WH Wilborn, JH Carey, R Ceballos, JW Benton, CA Alford: The simultaneous occurrence of subacute sclerosing panencephalitis in two brothers. I. clinical virologic and histopathologic studies. J. Infect. Dis. 125: 173-178, 1972.

36. Jabbour, JT, DA Duenas, JL Sever, HM Krebs, L Horta-Barbosa: Epidemiology of subacute sclerosing panencephalitis (SSPE). A report of the SSPE registry. J A M A 220: 959-962, 1972.

37. Macintyre, EH, JA Armstrong: Fine structural changes in human astrocyte carrier lines for measles virus. Nature 263: 232-234, 1976.

38. Lucas, A, M Coulter, R Anderson, S Dales, W Flintoff: In Vivo and In Vitro models of demyelinating diseases. II. Persistence and host-regulated thermosensitivity in cells of neural derivation infected with mouse hepatitis and measles viruses. Virology 88: 325-337, 1978.

39. Gibbs, FA, EL Gibbs, PR Carpenter, HW Spies: Electroencephalographic abnormality in 'uncomplicated' childhood diseases. J A M A 171: 1050-1055, 1959.

40. Chiarini, A, A Sinatra, P Ammatuna, R Distefano: Studies on a measles virus variant inducing persistent infections in cultured cells. I. Isolation and characterization of plaque purified virus clone. Arch. Virol. 52: 47-58, 1976.

41. Thormar, H, PD Mehta, HR Brown: Comparison of wild-type and subacute sclerosing panencephalitis strains of measles virus, Neurovirulence in ferrets and biological properties in cell cultures. J. exp. Med. 148: 674-691, 1978.

42. Follett, EAC, CR Pringle, TH Pennington, P Shirodaria: Events following the infection of enucleate cells with measles virus. J. gen. Virol. 32: 163-175, 1976.

43. Shirodaria, PV: In: Measles virus and its biology, Fraser and Martin (eds), London, Academic Press, 1977, p61.

44. Huang, AS, D Baltimore: Defective viral particles and viral disease processes. Nature (London) 226: 325-327, 1970.

45. Holland, JJ, LP Villarreal: Persistent non-cytocidal vesicular Stomatitis virus infections mediated by defective T particles that suppress virion transcriptase. Proc. nat. Acad. Sci. USA. 71: 2956-2960, 1974.

46. Rima, B, B Davidson, SJ Martin: The role of defective interfering particles in persistent infections of Vero cells by measles virus. J. gen. Virol. 35: 89-97, 1976.

47. Hall, WW, SJ Martin, E Gould: Defective interfering particles produced during the replication of measles virus. Med. Microbiol. Immunol. 160: 155-164, 1974.

48. Osunkoya, BO, AR Cooke, O Ayeni, TA Adejumo: Studies on leukocyte cultures in measles: I. Lymphocyte transformation and giant cell formation in leukocyte cultures from clinical cases of measles. Arch. Ges. Virusforsch. 44: 313-322, 1974.

49. Osunkoya, BO, GI Adeleye, TA Adejumo, LS Slimonu: Studies on leukocyte cultures in measles. II. Detection of measles virus antigen in human leukocytes. Arch. Ges. Virusforsch. 44: 323-329, 1974.

50. Joseph, BS, MBA Oldstone: Antibody-induced redistribution of measles antigen on the cell surface. J. Immunol. 113: 1705-1709, 1974.

51. Johnson, KP, P Swoveland: Measles antigen distribution in brains of chronically infected hamsters: An immunoperoxidase study of experimental subacute sclerosing panencephalitis. Lab. Invest. 37: 459-465, 1977.

52. Joseph, BS, MBA Oldstone: Immunologic injury in measles virus infection. II. Suppression of immune injury through antigenic modulation. J. exp. Med. 142: 864-876, 1975.

53. Janda, Z, E Norrby, H Marusyk: Neurotroprism of measles virus variants in hamsters. J. Infect. Dis. 124: 553-564, 1971.

54. Katz, M, H Koprowski: The significance of failure to isolate infectious viruses in cases of subacute sclerosing panencephalitis. Arch. Ges. Virusforsch. 41: 390-393, 1973.

55. Tijl, WFJ, SLH Notermans, M Katz, V ter Meulen, H Koprowsky: SSPE virus encephalitis produced experimentally in young dogs. Clinical and electro-encephalographic aspects. In: Proc. 13th Int. Congress Pediatrics Vienna, 3: 425-429, 1971.

56. Thein, P, A Mayr, V ter Meulen, H Koprowski, MY Kackell, D Müller, R Meyermann: Subacute sclerosing panencephalitis: Transmission of the virus to calves and lambs. Arch. neurol. 27: 540-548, 1972.

57. Notermans, SLH, WFJ Tijl, FTC Willems, JL Sloof: Experimentally-induced subacute sclerosing panencephalitis in young dogs. Neurology 23: 543-553, 1973.

58. Raine, CS, LA Feldman, RD Sheppard, MB Bornstein: Subacute sclerosing panencephalitis virus in cultures of organised central nervous tissue. Lab. Invest. 28: 627-640, 1973.

59. Iwaski, Y, H Koprowski: Cell to cell transmission of virus in the central nervous system. Lab. Invest. 31: 187-196, 1974.

60. ter Meulen, V, M Katz, YM Kackell: Properties of SSPE virus: tissue culture and animal studies. Ann. clin. Res. 5: 293-297, 1973.

61. Albrecht, P, AL Shabo, GC Barns, NM Tauraso: Experimental measles encephalitis in normal and cylophosphamide treated monkey. J. infect. Dis. 126: 154-161, 1972.

62. Byington, DP, KP Johnson: Subacute sclerosing panencephalitis virus in immunosuppressed adult hamsters. Lab. Invest. 32: 91-97, 1975.

63. Raine, CS, DP Byington, KP Johnson: Subacute sclerosing panencephalitis in the hamster: Illustrations of the acute disease in newborns and weanlings. Lab. Invest. 33: 108-116, 1975.

64. Byington, DP, KP Johnson: Experimental subacute sclerosing panencephalitis in the hamster: correlation of age with chronic inclusion cell encephalitis. J. infect. Dis. 126: 18-26. 1972.

65. Johnson, KP, EG Feldman, DP Byington: Effect of neonatal thymectomy on experimental subacute sclerosing panencephalitis in adult hamsters. Infect. Immun. 12: 1464-1469, 1975.

66. Kibler, R, NE Cremer: Anti-thymocyte treatment of hamsters inoculated with a measles virus extracted from a patient with subacute sclerosing panencephalitis. Immunol. Commun. 2: 303-321, 1973.

67. Albrecht, P, T Burnstein, MJ Klutch, JT Hicks, FA Ennis: Subacute sclerosing panencephalitis: experimental infection in primates. Science 195: 64-66, 1977.

68. Johnson, KP, E Norrby: Subacute sclerosing panencephalitis (SSPE) agent in hamsters. III. Induction of defective measles infection in hamster brain. Exp. mol. Path. 21: 166-178, 1974.

69. Carrigan, DR, RR McKendall, KP Johnson: CNS disease following dissemination of SSPE measles virus from intraperitoneal inoculation of suckling hamsters. J. med. Virol. 2: 347-357, 1978.

70. Ueda, S, T Otsuka, Y Okuno: Experimental subacute sclerosing panencephalitis (SSPE) in a monkey by subcutaneous inoculation with a defective SSPE virus. Biken J. 3: 179-181, 1975.

71. Waksman, BH, T Burnstein, RD Adams: Histologic study of the encephalomyelitis produced in hamsters by a neurotropic strain of measles. J. Neuropath. exp. Neurol. 21: 25-49, 1962.

72. Griffin, DE, J Mullinix, O Narayan, RT Johnson: Age dependence of viral expression: Comparative pathogenesis of two rodent-adapted strains of measles virus in mice. Infect. Immun. 9: 690-695, 1974.

73. Herndon, RM, L Rena-Descalzi, DE Griffin, PK Coyle: Age dependence of viral expression. Electron microscopic and immunoperoxidase studies of measles virus replication in mice. Lab. Invest. 33: 544-553, 1975.

74. Roos, RP, DE Griffin, RT Johnson: Determinants of measles virus (Hamster neurotropic strain) replication in mouse brain. J. infect. Dis. 137: 722-727, 1978.

75. Johnson, KP, DP Byington: Subacute sclerosis panencephalitis (SSPE) agent in hamsters. I. Acute giant cell encephalitis in newborn animals. Exp. Mol. Path. 15: 373-379, 1971.

76. Albrecht, P, HP Schumacher: Neurotropic properties of measles virus in hamsters and mice. J. infect. Dis. 124: 86-93, 1971.

77. Wear, DJ, F Rapp: Latent measles virus infection of the hamster central nervous system. J. Immunol. 107: 1593-1598, 1971.

78. Norrby, E, K Kristensson: Subacute encephalitis and hydrocephalus in hamsters caused by measles virus from persistently infected cell cultures. J. Med. Virol. 2: 305-317, 1978.

79. Schumacher, HP, P Albrecht, NM Tauroso: The effect of altered immune reactivity on experimental measles encephalitis in rats. Arch. Ges. Virusforsch. 37: 218-229, 1972.

80. Maugh, TH: Multiple sclerosis: two or more viruses may be involved. Science 195: 768-771, 1977.

81. Notkins, AL, SE Mergenhagen, RJ Howard: Effect of virus infections on the function of the immune system. Ann. Rev. Microbiol. 24: 525-538, 1970.

82. Haider, S, M de L Coutinho, RTD Edmond, RNP Sutton: Tuberculin anergy and infectious mononucleosis. Lancet 2: 74, 1973.

83. Mangi, RJ, MD Niederman, JE Kelleher, JM Dwyer, AS Evans, FS Kantor: Depression of cell-mediated immunity during acute infectious mononucleosis. New Eng. J. Med. 291: 1149-1153, 1974.

DISCUSSIONS OF PAPERS BY V. TER MEULEN AND K. B. FRASER

This discussion opened with some consideration of the role of the virus matrix protein. Attention then shifted to the epidemiology of SSPE. The high incidence in parts of the Middle East, particularly Turkey, was alluded to. The apparent absence of SSPE in Nigeria, where measles is common in early life, was disputed; Dr Cathala recalled that SSPE was, in fact, common in Nigeria where measles epidemics were followed by waves of SSPE. Although SSPE may occur in Kenya and Nigeria, there thus remained some doubt as to whether it was diagnosed as such. Probably, the very high figures quoted from some countries, (e.g. Turkey, Czechoslovakia and Hungary) related to the personal interest of the physician responsible for reporting. The data from the U.S.A. were regarded by Dr Lachmann as incredible and he was sceptical of the apparent increased prevalence in rural areas.

There was little protection by maternal antibody and it might well be that children who developed SSPE had an abnormal immunological response. The activation of other viruses and the development of antibodies to herpes simplex and mumps viruses in the CSF of patients with SSPE might indicate this. In the context of altered immuno- logical surveillance it was mentioned that cell-mediated immunity is commonly depressed in measles.

CANINE DISTEMPER ENCEPHALITIS

E. LUND

The distemper virus is very closely related to measles virus and the spectrum of canine and human diseases show remarkable similarities. In addition to the common acute disease there are occasional development of a perivascular demyelinating disease and the rare occurrence of chronic CNS
infection. In man this presumably may occur as subacute
sclerosing panencephalitis.

Canine distemper in dogs

Distemper in dogs has been studied for many years.
Some dogs develop an acute encephalitis during the late
course of disease, and in other dogs long term persistence
of virus occurs in CNS and lead to a chronic encephalitis
(old dog encephalitis). Gillespie and Richard pointed out
already in 1956 (4) that the basic factors influencing
distemper virus to produce encephalitis are poorly understood, and that a major difficulty is the failure to produce nervous manifestations regularly. They demonstrated
that a series of intracerebral inoculations in dog of a
strain, designated Snyder Hill, resulted in a material
that could produce neurological signs regularly. Raine
(15) reported a total of eight dogs which suffered between
10 days to 12 weeks of acute or chronic natural disease,
which could be segregated according to the lesions found

by histological manifestation. Viral inclusions were found
only within areas showing pathological changes, but active
multiplication of virus was not demonstrated. Perivascular
cuffing was frequently observed and edema. Eventually only
fibrous astrocytes remained. The pattern seemed similar to
the one of experimental allergic encephalomyelitis with
mononuclear cells invading the myelin sheath.

It has been reported (11) that the chance to get a
demyelinating process depends very much on the strain of
virus employed, and that an especially well suited strain
has been found. It has, however, not been available outside the group. The same group (9) has suggested that the
inability to produce antibodies to envelope antigens may
be a crucial factor in the establishment of a persistent
infection with canine distemper virus.

As pointed out by Johnson and Weiner (7) there has
for many years been controversy regarding the pathogenesis
of the acute demyelinating CNS disease, essentially
whether it is a direct viral effect or an immune-mediated
myelin destruction. It is not stated that after all both
mechanisms might be involved at the same time. They point
out that essentially no experimental model is available
to study the human post-infectious encephalomyelitis in the
laboratory. In a number of countries measles immunization
has successfully eliminated the disease in man, so that
clinical material is not available any more.

Møller (12) has performed post mortems on around 5000
dogs in the period before distemper vaccine of good quality was generally available. In the cases of acute distemper he demonstrated essentially inclusions in the cytoplasm. In the cases of chronic, i.e. old dog encephalitis,
he found intranuclear inclusions in cells of the central
nervous systems. It has been strongly suggested (1), that
there is a possible relationship of old dog encephalitis
to multiple sclerosis, subacute sclerosing panencephalitis
and neuromyelitis optica,and that old dog encephalitis is
a valuable model for further study of severe demyelinating
diseases of dogs and man. In the same publication it is,

however, admitted, that accurate diagnosis in the living
animal is still difficult, and that demyelination in old
dog encephalitis is usually diffuse.

Raine et al (16) described measles virus (Edmonston
strain) grown in central nervous system tissue, and how
the intranuclear presence of nucleocapsid became gradually
more abundant.

Koestner et al (8) have worked with the Lederle (vaccine) strain in explant cultures of canine cerebellum,
where they found inclusions and a number of viral changes.
As they can demonstrate demyelination in the cerebellum
cultures, they suggest the use of such cultures as an up-
propriate model for certain parts of the in vivo process.

It seems that the most significant study to explain
part of the pathogenesis is the one of Nakai, Shand and
Howatson (13). They described the development of measles
virus in vitro by electron microscopy and otherwise. They
found that nucleocapsids produced in the cytoplasm budded
off from the cellular membranes and became infectious virions, but that the nucleocapsids of the nuclei apparently
was a dead end not resulting in infectious virions. This
observation together with the intranuclear inclusions of
the old dog encephalitis seem potentially very important
as part of an hypothesis for measles-distemper encephalitis.

Distemper virus infection of mink

In all the literature dealing with measles and distemper encephalitis and the search for a relevant model for
the human infections it seems overlooked, that not only
Canines, but also Mustelidae may become infected with distemper virus, and that central nervous system involvment
in mink is at least as common as in canines. This has been
very obvious in Denmark, where epidemics of distemper have
occurred since commercial breeding started around 1940 and
became especially troublesome in the late 60'es. Mink farms
have been placed quite close, so that the density of mink
may be more than 100000 animals per 50 square kilometer,
and the yearly production is and has for several years been
several millions of animals. In the dense areas the morta-

lity of distemper could be up to 60-70 per cent. The mortality depends on a number of factors like genotypes of mink
and virus strain and very much on age. The kits are born
within a short period from late April and are except the
breeder animals pelted in December. In spite of this the
epidemics occurred very often in the autumn. Certain genotypes, like the pastels, had a higher morbidity and mortality than other genotypes, like the standards. Ferrets are
more sensitive than mink. The ability to form antibodies
does not depend on the genotypes (5, 6).

The epidemics varied in intensity, but also so that
the respiratory and conjunctivitis symptoms dominated in
some years and the hard-pad disease in other years. The
infection is very contagious. All animals become infected.
It seems that convulsions and other signs of CNS involvment were more frequent in years with hard-pad disease.
Convulsions may occur in animals which have been ill for
some time. They scream suddenly in a special high-pitched
way and die in a strong convulsion. By the end of an epidemic other animals have sometimes fatal convulsions without
any previous signs of disease. Animals which survive the
epidemics may much later die suddenly, e.g. during mating.
Fatal convulsions in connection with noise from low flying
airplanes have also been connected with a previous history
of a distemper epidemic. Although the Danish Fur Breeders
Association kept a very careful record of the number of distemper cases, there is unfortunately no information on the
frequency of nervous disorders in the epidemics.

A study of encephalitis caused by distemper in mink

With the purpose of studying the pathogenesis of distemper encephalitis in mink and possibly setting up an experimental model for the corresponding syndrom in man a
group in Copenhagen decided in 1976 to start a series of
experiments employing mink. We were aware of a number of
difficulties: [1] The lack of knowledge of the normal anatomy of the mink brain, [2] The surely quite different qua-
lities of the mink blood, especially compared to human

blood, [3] the limited resources, [4] the lack of a conveniently situated experimental farm, [5] the lack of an optimal strain of distemper virus.

It seems that the literature on the influence of immunological factors on distemper encephalitis contains quite conflicting points of view. We had expected that natural distemper would occur again and give us material to study, but that has not been the case.

The factors that might influence the development of neurological symptoms are the genotypes and histocompatibility type, the virus strain and dose, immunological factors and other infections like plasmacytosis ("Aleutian disease") and virus enteritis.

The decision was to look into the effect of immunological factors. We tried the treatments indicated in Table 1.

Table 1

Group	Treatment	Mortality in percentage
1	Immunosuppressive treatment employing cyclophosphamide starting one day before virus inoculation	35
2	Niridazol treatment starting one day before virus inoculation	70
3	Niridazol treatment starting 7 days after virus inoculation	75
4	Specific immunoglobulin starting one day before virus inoculation	72
5	Levamisol treatment starting one day before treatment	61
6	Levamisol treatment starting 7 days after virus inoculation	75
7	Heat inactivated vaccine	60

8	Proper vaccination employing live vaccines	O
9	Control receiving only virus inoculation.	30

Each group consisted of 20 animals, 4 months old and without pastel genes. They came from a farm that never had distemper outbreaks and thus were seronegative in neutralization test. Unfortunately a high number became plasmacytosis positive during the experiment. This could be avoided now through the better testing available.

ad 1) The broad, non-specific suppression of cyclophosphamide treatment has been valuable in connection with plasmacytosis antigen production in mink and has influence e.g. Marek disease in chickens (10).

ad 2 and 3) Niridazole, an anthelminthic drug, has been shown to be a potent, long-acting suppressant of cell-mediated immune responses (14).

ad 5 and 6) Levamisole (tetrahydro phenylimidazothiazole hydrochloride) is an anthelminthic drug, which apparently stimulates T-cell functions. It seems that the effect is functional especially in connection with subnormally functioning cells.

ad 7) We have previously seen (5) that such an antigen could cause formation of neutralizing antibodies, but not protection against distemper.

Three weeks after vaccination of group 5 and 6 all the animals were inoculated on Sept. 9 with a Snyder Hill strain of virus. The final result in terms of mortality was registered on Dec. 27 with the result given in Table 1. Only few (around eight) of the animals had signs of centralnervous system affection like ataxi, strange behaviour, excessive saliva, and the mortality of the control group was low. On 94 animals a preliminary post mortem was carried out immediately after death or at the end of the experiment. Very few cases of diffuse lesions in the brain with mononuclear cells perivascularly and no regular demyelination were observed. Plasmacytosis was partly inhi-

bitory for the proper evaluation. It was the intention to
make a more thorough examination of the material, but the
pathologist, professor Møller, died quite suddenly. In 18
mink some neurological changes were found. None of these
were animals of the control group or the vaccinated ones,
but the findings were perhaps not significant at all. Determination of neutralizing antibodies and a preliminary
study of leucocyte migration inhibition were carried out.
Virus isolations were carried out on throat and rectal
swabs and determinations of virus antigens by means of
indirect immunoperoxidase testing. No mink had demonstrable
virus excretion one week after inoculation, but by day 22
all animals were excreting virus, also the vaccinated ones.
No virus was demonstrated after 5 weeks. There was no correlation between antibody titre and course of infection.
Titres between 25 and 625 were found in all groups and
very few outside that range.

In the following year we performed a new study with
only two groups: Levamisol treated animals and control animals. The Snyder Hill strain was this time received from Behringwerke. The impression from previous
vaccination trials was that this strain gave a higher frequency of neurological symptoms than the strain employed
the year before.

The plan was to take out brain material for the
establishing of cell cultures and for histological examination from animals dying or with clear neurological symptoms. Leucocyte migration test (2) and lymphocyte cultures (3) for a blasttransformation test were attempted
in a number of modifications, but very often it proved to
be difficult to obtain a proper number of free white
blood cells, and no conclusions can be drawn from the results. Mink blood differs in many ways from that of other
species, and clumping and destruction of the desired cells
often occurred. In addition the wrong antigen, a vaccine
virus was used, which according to Schultz (17) would not
suppress the lymphocytes.

Distemper virus was demonstrated from most of the

cultivated brain tissues of the animals dying during the first 4 weeks after inoculation. In most cases the cells had to be transplanted to form new cultures, before cell degeneration and cytoplasmic inclusions could be seen.

The Levamisol treated animals had convulsions, sometimes fatal, more often than the controls (8/30 against 2/30). The animals died faster: 17 were left out of the 30 Levamisol treated on the 15th day after virus inoculation, but 22 out of the 30 controls. One week later 10 remained in both groups.

After 3 months the Levamisol treatment was repeated. In this case some animals died in convulsions soon after inoculation (1 ml of an 1/5 dilution given intramuscularly). Consequently it cannot be excluded that the substance was toxic to the animals, rather than enhanced the effect of the virus. The brain cultures set up after 3 months did not contain viral inclusions or show CPE. Electron microscopy did not reveal changes related to distemper virus.

Unfortunately no histological examinations have been carried out on the brain materials so far. Consequently it is not yet possible to draw any conclusions from the experiments.

In spite of a number of practical difficulties encountered with the experiments we have so far carried out, I would like to suggest that mink potentially are very useful animals for distemper experiments, much to be preferred to dogs.

REFERENCES

1. Adams, JM, WJ Brown, HD Snow, SD Lincoln, AW Sears, jr.,
   M Barenfus, TA Holliday, NE Cremer and EH Lennette:
   Old dog encephalitis and demyelinating diseases in man.
   Vet Pathol 12, 220-226, 1975.
2. Clausen, JE: Comparison between capillary tube and aga-
   rose migration technique in the study of human peripheral blood leukocytes. Acta Allergol 28, 145-158,
   1973.
3. Gerber, JD and AE Marron: Cell-mediated immunity and
   age at vaccination associated with measles inoculation and protection of dogs against canine distemper.
   Am J Vet Res 37, 133-138, 1976.
4. Gillespie,JH and CG Richard: Encephalitis in dogs produced by distemper virus. Am J Vet Res 17, 103-108,
   1956.
5. Hansen, M and E Lund: The protecting capacity of neutralizing antibodies by distemper virus infections in
   mink. Acta path microbiol scand 80, 795-800, 1972.
6. Hansen, M and E Lund: Comparative distemper vaccine titrations in ferrets and mink. Nord Vet-Med 25, 1-7,
   1973.
7. Johnson, RT and LP Weiner: The role of viral infections
   in demyelinating diseases In: Multiple Sclerosis,
   Wolfgram, F, GW Ellison, JG Stevens and JM Andrews
   (eds), UCLA Forum in Medical Sciences No 16, New York
      AP 1972, p 245-264.
8. Koestner, A and JF Long: Ultrastructure of canine distemper virus in explant tissue cultures of canine cerebellum. Lab Invest 23, 196-201, 1970.

9. Krakowka, S, R Olsen, A Confer, A Koestner and B McCullough: Serologic response to canine distemper viral antigens in gnotobiotic dogs infected with canine distemper virus. J Infect Dis 132, 384-393,1976.

10. Lu, Y-Sh, V Kermani-Arab and T Moll: Cyclophosphamide-induced amelioration of Marek's disease in Marek's susceptible chickens. Am J Vet Res 37, 687-692, 1976.

11. McCullough, B, S Krakowka, A Koestner and J Shaddock: Demyelinating activity of canine distemper virus isolates in gnotobiotic dogs. J Infect Dis 130, 343-350, 1974.

12. Møller, T, personal communications.

13. Nakai, T, FL Shand and AF Howatson: Development of measles virus in vitro. Virology 38, 50-67, 1969.

14. Pelley, RP, RJ Pelley, AB Stavitsky, AAF Mahmoud and KS Warren: Niridazole, a potent long-acting suppressant of cellular hypersensitivity. J Immun 115, 1477-1482, 1975.

15. Raine, CS: Viral infections of nervous tissue and their relevance to multiple sclerosis. In: Multiple Sclerosis, Wolfgram, F, GW Ellison, JG Stevens and JM Andrews (eds), UCLA Forum in Medical Sciences No 16, New York, AP 1972, p 91-118.

16. Raine, CS, LA Feldman, RD Sheppard and MB Bornstein: Ultrastructure of measles virus in cultures of hamster cerebellum. J Virology 4, 169-181, 1969.

17. Schultz: Failure of attenuated canine distemper virus (Rockborn strain) to suppress lymphocyte blastogenesis in dogs. Cornell Vet 66, 27-31, 1976.

DISCUSSIONS OF PAPER BY E. LUND

The provision of suitable animal models for the study of SSPE provoked
much comment. Thus, in hamsters and ferrets, an SSPE-like condition
can be induced by the intracerebral inoculation of non-productive
virus. Doubts were expressed as to whether distemper in dogs or in
mink resembled SSPE clinically; Dr Lund replied that illnesses
following an acute course and also a more prolonged course were found
in the natural disease. Passage of the presumably infectious agent
in 'old-dog encephalitis', where inclusions are found in brain tissue,
had not been successful.

Dr Norrby summarised the model systems available and considered
that the best one consisted of the inoculation of cell-associated
measles-SSPE virus into ferrets. In hamsters, older animals (3 weeks
of age) could be used. In newborn hamsters, inoculation with
defective ts virus mutants from persistently infected tissue cultures
led to persistent infections. In the rat model, the newborn animals
had maternal antibody.

THE IMMUNE RESPONSE IN SUBACUTE SCLEROSING PANENCEPHALITIS

P. J. LACHMANN, N. T. GORMAN, J. HABICHT, P. W. EWAN, G. AGNARSDOTTIR AND H. VALDIMARSSON

1. INTRODUCTION

The idea that some rare, and not obviously contagious, human diseases may result from an unusual immunological response to commonly encountered infectious agents has attracted increasing attention in recent years. While much of this interest was stimulated by studies in mice, for example on chronic LCM disease (1) it is nevertheless probable that the clearest example of a disease where this view of pathogenesis is correct is subacute sclerosing panencephalitis (SSPE). The disease is extremely rare while the virus causing it is virtually universally encountered. Thus, in the United States before the advent of immunisation against measles it was estimated that the incidence of SSPE was one case for every million cases of measles although in the Middle East it seems to be significantly commoner (see 2). The disease is not contagious in as much as there are no reports of anyone having caught SSPE or even measles from an SSPE patient. Finally, there is no doubt that the immunological response to the measles virus in SSPE is unusual and it is the nature of this response which is the topic of this paper.

2. THE IMMUNE RESPONSE TO MEASLES VIRUS IN SSPE

Measles virus contains six major polypeptides (3,4,5). These are shown with their molecular weights (and with the comparable proteins of canine distemper virus) in Table 1. By using immunoprecipitation techniques followed by polyacrylamide gel electrophoresis (PAGE) analysis it is possible to detect in serum, antibodies to all six of these polypeptides (6). However, this has been relatively infrequently done and most of the work on the antibody response to measles virus has used other techniques: haemagglutination inhibition which measures antibodies to the haemagglutinin; haemolysis inhibition which measures antibodies to the fusion factor or haemolysin; neutralisation which may measure both the two previous antigens; and

Table 1   Comparison between measles and canine distemper viral
           proteins

			M.V.	C.D.V.	Present on membrane of infected cell
L	–	Large	200K	200K	O
H	–	Haemagglutinin	80K	76K	+
P	–	Phosphorylated	70K	66K	O
NP	–	Nucleocapsid	62K	58K	O
F_o	–	Fusion factor or haemolysin	55K	60K	+
M	–	Matrix protein	37K	34K	??
F_o	=	$F_1 + F_2$ (on reduction)			
		F_1	40K	40K	+
		F_2	15K	20K	+

complement fixation which measures antibodies principally to the nucleocapsid. In considering the pathogenesis of virus infections, particular interest attaches to antibodies towards those viral proteins that are represented on the membrane of the infected cell. In the case of the measles virus these are the haemagglutinin and fusion factor. The possibility that the matrix protein may be present on the membrane cannot be wholly excluded (7) though in our experiments we have not been able to detect it on cell membranes by the technique of cell membrane radio-labelling followed by immuno-precipitation and PAGE analysis (8).

2.1. The antibody response to the measles virus in SSPE

Connolly et al (9) first reported that patients with SSPE had very high levels of antibodies to measles virus both in serum and in CSF. This has been borne out in all subsequent studies. The levels can be really extremely high. Thus 10 - 20% of the IgG in one of the sera that we have studied could be bound to the surface of pers-istently measles virus infected cells. This measures only antibody to the haemagglutinin (H) and the fusion factor (F) and it therefore seems likely that probably as much as 20% of the IgG in this patients serum may have been antibodies to the measles virus. This sort of antibody level is comparable to what can be achieved in a hyper-immunised rabbit! The high levels of measles antibody can be detect-ed by any of the standard techniques as well as by immunoprecipita-tion and PAGE analysis. Using the last named technique Hall et al (6) have claimed that antibody titres in SSPE are raised to all the measles virus polypeptides except one. Antibodies to the matrix protein were found at only low levels in a number of SSPE sera that they studied. This selective relative failure to make antibody to the M protein was characteristic of SSPE sera in their study. They suggest that this may reflect a failure of measles virus in SSPE to produce adequate a mounts of M protein and they believe that this may form an integral part of the pathogenesis of the disease.

Besides having high levels of anti-measles antibodies in their sera patients with SSPE also show high levels in the CSF. The anti-body response in the CSF is characteristically oligoclonal and it has been shown by Vandvik et al (10) that individual oligoclonal bands

within the CSF represent individual measles virus antigens. Oligoclonal Ig bands in the CSF are probably a marker of antibody that is
synthesised within the CNS and it is believed that this reflects the
fact that the number of precursor cells within the CNS is small, and
that this does not allow the development of the high degree of poly-
clonality that is seen in the serum when antibody synthesis is initiated by a much larger number of precursor cells.

2.1.1. Antibody affinity

There have been no substantive studies on the affinity of the anti-
measles antibodies in SSPE. Some preliminary studies were carried
out some years ago (Lachmann & Fazekas de St. Groth unpublished
observations) which appeared to show that in the SSPE sera there was
a mixture of low affinity with a smaller concentration of antibodies
of a high affinity.

2.2 Immunoglobulin and antibody synthesis in the CNS as measured by the Tourtellotte technique

Tourtellotte (12, 13) described a relatively non-invasive method for
determining the extent to which immunoglobulin is synthesised within
the CNS. This depends upon the simultaneous measurement of albumin
and IgG levels in serum and CSF. From the albumin levels an expected
value of the IgG level can be calculated, this being made up of Ig
that is transuded across a normal blood brain barrier plus a quantity
that is exuded across a leaky blood brain barrier. Although there
is some doubt about the assumptions underlying the calculation of the
exuded IgG (see 11) and the confidence intervals of the calculations
are therefore wider in patients with leaky blood brain barriers,
there is no doubt that a reasonable estimate of the IgG synthesised
in the CNS can be obtained by subtracting from the observed IgG level
that calculated from the albumin levels as derived from plasma. We
have used the Tourtellotte technique for measuring IgG synthesis in
patients with SSPE as well as with multiple sclerosis (11,14) and the
results are shown in Figure 1. Here it can be seen that the IgG synthesis levels in SSPE are remarkably high even compared to those found
in MS.

The principle underlying the Tourtellotte technique is of
course equally applicable to the measurement of individual antibodies
providing that these can be accurately quantitatively measured in this
way (14, 15). In order to obtain an adequately quantitative

Figure 1

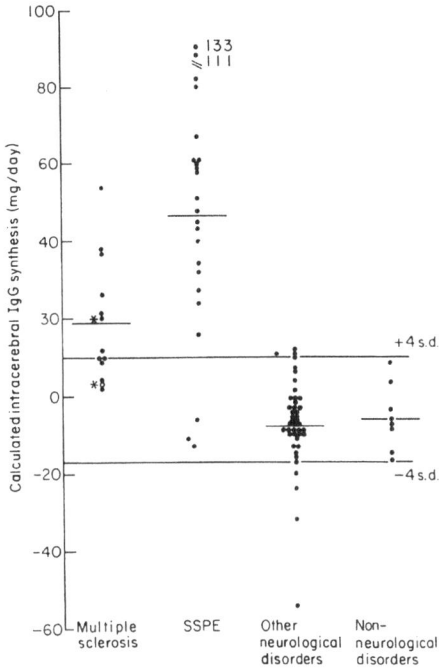

Intracerebral IgG synthesis in MS, SSPE, and two control groups. * Indicates results in one MS patient studied on two occasions. The normal range indicated (mean ± 4 s.d.) is taken from Tourtellotte's data obtained in normal subjects. Horizontal bars indicate mean for each group. The differences between MS and the neurological control group, and SSPE and the neurological control group are statistically significant ($P < 0.001$ for both by Student's t-test).

From Ewan & Lachmann (1979)

assay the antibodies were measured by competitive radioimmunoassay
using as antigen fixed HeLa cells persistently infected with measles
virus. This assay is likely to measure predominantly antibodies to
haemagglutinin and to the fusion factor. There is a problem in the
use of competitive radioimmunoassay for this purpose since this
measures the antibody in inhibitory units compared to the standard
preparation of SSPE serum and since the affinity of the antibody as
well as its concentration plays a part in this inhibitory activity
the values obtained will contain errors if the affinity of antibody
in serum and CSF are not the same. Table 2 shows the results of the
measles antibody synthesis experiments in a group of SSPE patients
compared with two groups of MS patients: one showing elevated IgG
synthesis and the other not so doing and with a group of normal
patients. It can be seen that there is indeed a very marked hyper-
synthesis of anti-measles virus antibody within the CNS. The amount
detected by this technique was about 20% of the total IgG synthesis
as measured. Since there are antibodies to at least several other
measles virus polypeptides that are not detected by this assay these
findings are compatible with the great majority of the antibody in
the CNS being anti-measles virus. The same is not true for the MS
patients where the amount of anti-measles IgG synthesis found is
independent of whether or not any overall IgG synthesis is found and
where it is little more than is found in a control group. It is
interesting that low levels of antibody synthesis are detected by
this technique even when no overall IgG synthesis can be found. It
remains unclear whether this is an artefact of the measuring technique
and depends on the fact that the antibody formed in the CNS is of
particularly high affinity. It is certainly true, however, that the
ratio of antibody between serum and CSF is much lower than is found
for the total IgG level and that in our hands it is not markedly
different in SSPE from the ratio found in normals or in patients with
MS. This finding was unexpected since it is at variance with what
has previously been published (16). However it is possible to use
this data to make a rough calculation of what proportion of the body's
anti-measles antibody is being made within the CNS. These calculations are shown in Table 2, and it can be seen that the 8% found in
SSPE is not significantly higher than in normals. This shows that

Table 2   Measles virus antibody synthesis in CNS

	SSPE	MS(1) +ve IgG synth.	MS(2) −ve IgG synth.	Other
Number	4	6	4	6
Blood brain barrier	292	290	184	273
I/c IgG synthesis mg/day	31.8	16.4	−2.3	−8.4
Measles antibody levels				
Serum μg/ml	529	27	15.4	16.5
CSF μg/ml	13.4	0.9	1.1	0.5
Ratio serum/CSF	39.5	30.3	14.7	33
I/c measles antibody synthesis mg/day	6	0.4	0.5	0.2
I/c measles antibody synthesis as % of i/c IgG synthesis	18.9%	2.3%	−	−
I/c measles antibody synthesis as % of total measles antibody synthesis in body	8%	5%	12%	6%
Total measles antibody synthesis in body (approx.) mg/day *	101	6	4	4
I/c measles antibody synthesis as % of total measles antibody synthesis in body	8%	5%	12%	6%

*Calculation assumes plasma volume of 5 litres and further ECF volume of 10 litres (with antibody concentration one half that of plasma)for adults;(for children with SSPE 80% of these volumes were assumed); and assumes half life of IgG to be 21 days.

Then total anti-measles IgG(mg) = 10 x plasma conc. (μg/ml)

and synthesis rate in mg/day $= \dfrac{\text{total Ab}}{42}$

$$= \dfrac{\text{plasma conc. (μg/ml)}}{4.2}$$

although there is undoubted and considerable hypersynthesis of anti-measles antibody (about 6mg/day compared with 0.2mg/day in normals) within the CNS there is an entirely comparable degree of hyper-synthesis in the rest of the body (about 100mg/day cf 4mg in normals). The amount of these particular anti-measles virus antibodies made is much greater than the total intracerebral IgG synthesis and it is therefore impossible that a large proportion of it could be made within the CNS. The degree of systemic hyperreactivity in SSPE is not generally taken full account of in discussing theories of the pathogenesis of the disease. The abnormality is not situated wholly within the brain although persistent leakage of antigen from CNS into the systemic circulation may be responsible for the chronic hyper-immunisation. The data also shows that ratios of serum/CSF antibody titres are an unsatisfactory way of showing intracerebral synthesis.

Comparable studies have been done using distemper virus (14). This was done largely in view of the interest in the possibility that distemper may be implicated in the pathogenesis of multiple sclerosis, but the SSPE patients are an interesting control group for this purpose since it is clear that their disease is due to the measles virus. The distemper data are shown in Table 3. It can be seen that the levels of antibody found are very much lower than those detected against the measles virus and the somewhat elevated level in SSPE is probably a sign of cross reactivity. It is again of interest how low the ratios of serum to CSF antibody are for this virus in all groups of patients. In fact the lowest values are found among the groups of normals and those MS patients showing no IgG synthesis. Again, taking all the reservations about these techniques into account it would appear that a substantial proport-ion of anti-distemper antibody found is made within the central nervous system in normals. The ratio of measles antibody to distem-per antibody in both serum and CSF is much the same in normals and in both MS groups but is much elevated in SSPE as might be expected. These findings lend little support to the idea that there is a significant immune response to the distemper virus in multiple sclerosis but they do show that the techniques can distinguish the measles related response in SSPE from a distemper virus cross-reacting response.

Table 3   Distemper virus antibody synthesis in CNS

	SSPE	MS(1) +ve IgG synth.	MS(2) −ve IgG synth.	Other
Number	4	6	4	6
Blood brain barrier	292	290	184	273
IgG synthesis mg/day	31.8	16.4	−2.3	−8.4
Distemper antibody levels				
Serum µg/ml	3.6	1.6	1.0	0.8
CSF µg/ml	0.2	0.1	0.3	0.1
Ratio serum/CSF	18	12.2	3.9	5.4
I/c distemper antibody synthesis mg/day	.13	.05	.1	.05
I/c distemper antibody synthesis as % of i/c IgG synthesis	0.4%	0.3%	−	−
Total distemper antibody synthesis in body (approx.) mg/day	.7	.4	.2	.2
I/c distemper antibody synthesis as % of total distemper antibody synthesis in body*	25%	11%	47%	23%

* As in Table 2

2.3. The cellular immune response to the measles virus in SSPE

There are considerable problems attending the studies of cell-
mediated immune responses in patients with measles and to the measles
virus in particular. These are related firstly to the well known
property of the measles virus in producing anergy in delayed hypersensitivity reactions (17). Furthermore the virus binds to human
T lymphocytes, not by specific anti-measles virus receptors but by
some receptor on the T cell as a class that has affinity for the
measles virus (18). There are technical problems with lymphocyte
transformation tests using measles virus preparations as antigen and
the techniques that have been mainly used are (i) delayed hypersensitivity skin tests (ii) the production of lymphokines usually
macrophage migration inhibitory factor and (iii) cytotoxic reactions
killing measles virus infected cells. In vivo skin testing with
measles virus preparations is almost invariably negative in patients
with SSPE (19, 20). The in vitro tests using migration inhibition
are, however, generally positive providing that they are done in the
presence of normal and not SSPE serum (21). This blocking effect of
SSPE which can be demonstrated both on lymphokine production and in
cytotoxicity assays is believed to be due to immune complexes (22)
although these have not so far been adequately studied by modern
techniques.

2.3.1 Cytotoxic reactions using measles virus infected cells
This topic will be treated only briefly since it is to be discussed
by Dr. Kreth in the next paper. The mechanism by which lymphocytes
kill measles virus infected cells remains a matter of some controversy. The reason is almost certainly that a number of mechanisms
can come into play (Table 4) and that the extent to which each one
is detected depends upon the exact experimental system used. The
antigens expressed on the cell membrane are likely to be important
and there is reason to believe that acutely infected cells are better
targets at least for some cytotoxic mechanisms than are persistently
infected cells (8).

It was shown by Valdimarsson et al (18) that embyronic lung
cells persistently infected with measles virus could be killed by
lymphocytes in the absence of antibody. Such killing was given even
by the lymphocytes of children who had never encountered the measles

Table 4   Mechanisms for killing measles virus infected cells

	Killer cell	Target cell			HLA preference		Found by
	Fc receptor	From immune donor	Infected with virus	Needs antibody			
1	+	0 (normal in SSPE)	+	+	0	K cell	All groups
2	0	0 or +	+	0	0	'NK' cell	Valdimarsson(1975)
3	0	+ (?low in SSPE)	+ (acutely inf.)	0 (typically inhibits)	0	? T cell ? NK cell	Ewan & Lachmann (1977; 1979)
c.p.	+	+ Mumps virus + only to induce cyto-toxic cell			0	NK cell	Harfast et al(1978)
4	0	+ (acute measles only)	+	0	+	T cell	Kreth et al(1979)

virus although the cells of adults who had encountered the virus
gave stronger killing. This 'non-allergic cell-mediated killing'
given by the cells of measles-negative children would presumably now
be referred to as NK killing. NK killing of virus infected cells
has been studied by Harfast et al (23) using mumps virus who showed
that the role of the virus is to activate the effector cell which can
then kill non-specifically. The effector cell in the studies of
Harfast et al is an Fc positive lymphocyte showing T cell markers.
It may be that in the case of the measles virus where attachment of
T cells occurs spontaneously that cytotoxicity may be given by a
wider selection of T lymphocytes. The killing of measles virus infected cells by the K cell mechanism is generally agreed to occur
(24,25,26,27). It requires the presence of antibody on the infected
cell and is mediated by an Fc receptor positive lymphocyte. All
groups also agree that K cell activity is normal in SSPE. Although
easily demonstrable in vitro it remains a question of some doubt
whether this mechanism is effective in vivo since it is inhibited
not only by physiological concentrations of immunoglobulin but is
readily inhibited by small concentrations of immune complex. A third
type of cytotoxicity to measles virus infected cells was described by
Ewan & Lachmann (27). This was mediated by populations of lymphocytes depleted of Fc receptored cells - a population which is largely
T cell in nature - and was not potentiated by the presence of antibody on the cell. In fact it is typically inhibited by anti-measles
antibodies in the test system. The killing produced by these cells
does not show HLA preference and the killing is now known to occur
much more readily on acutely infected than on persistently infected
cells (8). This type of killing is found variably among normal adults
and is present generally only weakly in patients with multiple
sclerosis. The two patients with SSPE who were tested also showed
low levels of this type of killing. It remains to be shown whether
this type of T cell killing is mediated by the T cell antigen binding
receptor or whether it is a variant of the NK type of cytotoxicity.
T cell cytotoxicity which does show HLA preference has been shown by
Kreth et al (28) to be demonstrable if the cells are taken from
children with acute measles. This form of killing is therefore more
likely to represent T cell killing through its antigen binding
receptor. There seems to be no clear evidence of any defect in
SSPE patients in their capacity to kill measles virus infected cells.

The antibodies in SSPE serum are also highly effective in sensitis-
ing measles virus infected cells for K cell killing. There is
therefore no reason to believe that the ability to kill measles virus
infected cells is intrinsically abnormal in SSPE. The extent to
which such killing could be inhibited by immune complexes is
difficult to assess. There have been some demonstrations of blocking
factors in serum and CSF (18, 22) but also others who have failed to
show it (24, 26). The amount of antigen expressed on the target
cell appears to influence how easily the reaction is inhibited (20).
Neither the serum nor even the lumbar CSF however, is wholly
representative of the microenvironment of the neurone where it seems
quite likely that immune complexes are being formed. Immune complexes involving measles virus have been detected in the kidney (29) so
they presumably circulate in the systemic circulation from time to
time.

3.   IMMUNE RESPONSE TO ANTIGENS OTHER THAN MEASLES VIRUS

The other prominent immunological abnormality in SSPE especially in
advanced cases is anergy on skin testing with a variety of other
antigens, for example DNCB, streptokinase-streptodornase, candida
and tuberculin (19). This, it is believed, reflected persistent
infection with the measles virus. It is interesting that lymphocyte transformation tests to the same antigens are usually normal.
The cutaneous anergy is relatively readily reversible by the administration of transfer factor (19) though it is very doubtful whether
this treatment has any effect upon the progress of the disease.
Other parameters of the immune response in SSPE (e.g. immunoglobulin
levels and lymphocyte populations) have generally been found to be
normal. It has recently, however, been found by H. Valdimarsson
& G. Agnarsdottir (personal communication) that in rapidly progressive SSPE there may be some reduction in the number of circulating T
cells as measured by rosetting with sheep erythrocytes.

3.1. HLA association

In all diseases with a putative immunological pathogenesis associat-
ions with particular HLA antigens are looked for. In SSPE these have
so far been unsuccessful. There was one claim (30) that HLA W29 was
associated with the disease but this association was not maintained

when large numbers of patients were studied (2). The discordant
occurrence of SSPE in identical twins (who presumably caught their
measles at the same time) also make it unlikely that there is a major
genetic determinant in the disease. However two pairs of siblings
with SSPE have been seen in London in the last five years.

4.   DISCUSSION

The immune response in children with SSPE is dominated by their high
level of antibodies to measles virus protein. With the exception of
the report by Hall et al (6) that there is a relative failure to
make antibody to the matrix protein this hyperreactivity extends to
all the polypeptides in the measles virus and the antibodies have not
been shown to have any differences in specificity from those found in
patients convalescent from measles.

   It was suggested many years ago that SSPE might represent an
example of immune deviation there being a failure of T cell responsiveness to the measles virus with corresponding hyperreactivity of
the B cell system. Put into more modern terminology one might say
that a failure of T suppressor cells allows an excessive antibody
response. There is so far no good data on suppressor cell function
in SSPE but if there is a defect it must be specifically towards
the measles virus since other antibody titres are not obviously
abnormal and nor indeed is there any clear evidence of a failure of
T cell immunity even towards the measles virus. This subject is
however complicated by the capacity of measles virus infection to
produce anergy and skin tests to a variety of antigens are negative.
On the other hand in vitro tests, such as they are, usually fail to
show any marked abnormality in the SSPE children. It is by no means
clear whether the persistence of the measles virus infection in the
brain is a consequence of the abnormal immunological response to the
virus; or whether the abnormal immunological response to the virus
is due to the persistence of the infection of the brain; or whether
the two interact in a cyclical fashion. Perhaps the best suggestion
for the pathogenesis of SSPE remains that of Joseph & Oldstone (31)
who showed that when measles virus infected cells are grown in the
presence of antibody but in the absence of an intact complement
system the cells become modulated so that the surface antigens are no
longer expressed on the cell membrane. The cells are then no longer

able to be destroyed either by antibody and complement or by lymphocytes and the virus is effectively sequestered intracellularly. In
such a situation one can envisage the virus infection persisting,
occasionally releasing virions (complete or incomplete) or occasionally spreading to other cells by direct cell fusion.

What factors allow modulation and persistence rather than
elimination of the virus remains obscure. It is unlikely to be only
lack of complement since SSPE has never been described in association
with genetic complement deficiencies.

5. SUMMARY

The outstanding feature of the immune response in SSPE is the greatly
exaggerated antibody response to the measles virus. This is found,
to a similar extent, both within the CNS and elsewhere in the body.
Antibody is made to all the measles virus antigens but it is claimed
that antibodies to the matrix protein are under-represented. There
is no convincingly demonstrated failure of T cell reactivity.

Whether hyperimmunisation with measles virus is the cause of
the persistent infection or its result remains to be determined.

REFERENCES

1. Oldstone, MBA, FJ Dixon: Pathogenesis of chronic disease assoc-
    iated with persistent lymphocytic choriomeningitis viral
    infection. J. Exp. Med. 129:483-505, 1969.

2. Lachmann,PJ, H Valdimarsson, G Agnarsdottir: The aetiology of
    subacute sclerosing panencephalitis. Immunopathology VII,
    ed. PA Miescher, Schwabe, Basel p. 148, 1977.

3. Wechsler, SL, BN Fields: Intracellular synthesis of measles virus
    specified polypeptides. J. Virol. 25: 285-297, 1978.

4. Graves, MC, SM Silver, PW Choppin: Measles virus polypeptide
    synthesis in infected cells. Virology 86: 254-263, 1978.

5. Tyrell, DLJ, E Norrby: Structural polypeptides of measles virus.
    J. Gen. Virol. 39: 219-229, 1978.

6. Hall, WW, RA Lamb, PW Choppin: Measles and SSPE virus proteins:
    Lack of antibodies to the M protein in patients with subacute
    sclerosing panencephalitis

7. Fraser, KB, M Gharpure, PV Shirodania, MA Armstrong, A Mave,
    E Dermott: In: Negative Strand Viruses and the Host Cell.
    Eds. BWJ Mahy, RD Barry, Academic press, pp.771-780, 1978.

8. Ewan, PW, PJ Lachmann: in preparation, 1979

9. Connolly, JH, I Allen, IJ Hurwitz, JHD Millar: Measles virus
    antibody and antigen in subacute sclerosing panencephalitis.
    Lancet 1: 542-544, 1967.

10. Vandvik, B, E Norrby, HJ Nordal, M Degre: Oligoclonal measles
    virus specific IgG antibodies isolated from cerebrospinal
    fluids, brain extracts and sera from patients with subacute
    sclerosing panencephalitis and multiple sclerosis. Scand. J.
    Immunol.5: 979-992, 1976.

11. Ewan, P, PJ Lachmann: IgG synthesis within the brain in multiple
    sclerosis and subacute sclerosing panencephalitis. Clin. exp.
    immunol. 35: 227-235, 1979.

12. Tourtellotte, WW: On cerebrospinal fluid immunoglobulin-G (IgG)
    quotients in multiple sclerosis and other diseases. A review
    and a new formula to estimate the amount of IgG synthesised
    per day by the central nervous system. J. Neurol. Sci. 10:
    279-304, 1970.

13. Tourtellotte, WW: What is multiple sclerosis? Laboratory
    criteria for diagnosis. Multiple sclerosis Research ed.
    AN Davison, JH Humphrey, AL Liversedge, WI McDonald, JS
    Porterfield. H.M.S.O. Lond. 1975.

14. Gorman, NT, PJ Lachmann, J Habicht: in preparation, 1979.

15. Lachmann, PJ, J Habicht: Synthesis of IgG and of anti-measles
    antibodies within the central nervous system in patients with
    subacute sclerosing panencephalitis and with multiple sclerosis.
    In: Progress in Neurological Research. Eds. P Behan, F Clifford
    Rose, Pitman, in press.

16. Norrby, E, H Link, JE Olsson: Measles virus antibodies in multiple
    sclerosis. Arch. Neurol. (Chicago) 30: 285-292, 1974.

17. von Pirquet, C: Das Verhalten der kutanen Tuberkulinreaktion
    während der masern. Dtsch.med.Wschr. 34: 1297, 1908.

18. Valdimarsson, H, G Agnarsdottir, PJ Lachmann: Measles virus
    receptor on human T-lymphocytes. Nature 255, 554-556, 1975.

19. Valdimarsson, H, G Agnarsdottir, PJ Lachmann: Cellular immunity
    in subacute sclerosing panencephalitis. Proc. Roy. Soc.
    Med. 67: 1125-1129, 1974.

20. Valdimarsson, H, G Agnarsdottir, PJ Lachmann: Subacute sclerosing
    panencephalitis. In: Clinical Neuroimmunology. Ed. F. Clifford
    Rose, Blackwell Scientific Publications, pp. 406-418, 1979.

21. Thurman, GB, A Ahmed, DM Strong, RC Knudsen, WR Grace, KW Sell:
    Lymphocyte activation in subacute sclerosing panencephalitis
    virus and cytomegalovirs infections. J. Exp. Med. 138: 839-846,
    1973.

22. Ahmed, A, DM Strong, KW Sell, GB Thurman, RC Knudsen, R Wistar,
    WR Grace: Demonstration of a blocking factor in the plasma and
    and spinal fluids of patients with subacute sclerosing pan-
    encephalitis. J. Exp. Med. 139:902-924, 1974.

23. Härfast, B, T Andersson, P Perlmann: Immunoglobulin-independent
    natural cytotoxicity of Fc receptor-bearing human blood
    lymphocytes to mumps virus-infected target cells. J. Immunol.
    121: 755-761, 1978.

24. Perrin, LH, A Tishon, MBA Oldstone: Immunologic injury in measles
    virus infection. III. Presence and carahcterisation of human
    cytotoxic lymphocytes. J. Immunol. 118: 282-290, 1977.

25. Kreth, HW, V ter Meulen: Cell-mediated cytotoxicity against measles
    virus in SSPE. 1. Enhancement by antibody. J. Immunol. 118:

291-295, 1977.

26. Kreth, HW, G Wiegand : Cell-mediated cytotoxicity against measles virus in SSPE. II. Analysis of cytotoxic effector cells. J. Immunol. 118: 296-301, 1977.

27. Ewan, PW PJ Lachmann: Demonstration of T-cell and K-cell cytotoxicity against measles-infected cells in normal subjects, multiple sclerosis and subacute sclerosing panencephalitis. Clin. exp. immunol. 30: 22-31, 1977.

28. Kreth, HW, V ter Meulen, G Eckert: Demonstration of HLA restricted killer cells in patients with acute measles. Med. Microbiol. Immunol. 165: 203-214, 1979.

29. Dayan, AD, MI Stokes: Immune complexes and visceral deposits of measles antigens in subacute sclerosing panencephalitis. Brit. Med. J. ii: 374-376, 1972.

30. Kurent, JE, JL Sever, PI Terasaki: HLA W29 and subacute sclerosing panencephalitis. Lancet 1: 927-928, 1975.

31. Joseph, BS, MBA Oldstone: Antibody induced redistribution of measles virus antigens on the cell surface. J. Immunol. 113: 1205-1209, 1974.

RECENT FINDINGS ON CELL-MEDIATED IMMUNE REACTIONS IN ACUTE MEASLES AND SSPE

W. KRETH AND F. PABST

1. INTRODUCTION

It is now more than 10 years that BURNET published his hypothesis on the pathogenesis of SSPE (1). BURNET assumed that the underlying defect was a specific unresponsiveness at the level of T cells. Despite a large body of experimental data (reviewed in (2)) this hypothesis cannot be conclusively answered at present. It is also evident that the mode of action of specific T cells during acute measles itself is incompletely understood. This may be at least partly due to the difficulties in designing appropriate in vitro experiments which explore conclusively specific T cell immunity in man.

2. HETEROGENEITY OF T CELL EFFECTOR FUNCTIONS

Most of the information on T cells has been derived from experiments in mice. By means of antigenic and functional markers 4 different subpopulations of T cells can now be recognized in mice (3): T helper cells (T^H), cytotoxic T lymphocytes (T^{CTL}), T cells involved in delayed type hypersensitivity (T^{DTH}), and T cells endowed with regulatory functions.

In SSPE, a deficient T cell response against measles antigens could extend to all T cell subpopulations. However, this is rather unlikely. Patients with SSPE produce extremely high amounts of specific antibody of IgG type (2). Also, the response to paramyxoviruses has been shown to be T cell-dependent in nude mice (4). Both lines of evidence strongly suggest that specific T helper cells must be functionally unimpaired in patients with SSPE.

As has been demonstrated in animal models, cytotoxic

T lymphocytes might play a major role in elimination of
host cells with altered surface moieties, such as tumor
and virus-infected cells. This was the reason why over
the last 5 years many investigators concentrated on
lymphocyte-mediated cytotoxicity against measles virus-
infected target cells in SSPE (5,6,7,8,9). Peripheral
SSPE lymphocytes were indeed found to be effective killer cells. Since it was assumed that these killer cells
belonged to the T cell lineage it was generally concluded that specific T cell mediated immunity was intact in
patients with SSPE. Today these conclusions can no
longer be maintained. It is the scope of this communication to critically analyze lymphocyte-mediated killing
in acute measles and SSPE.

3. SPONTANEOUS LYMPHOCYTE-MEDIATED CYTOTOXICITY IN SSPE:
INVOLVEMENT OF NATURAL KILLER CELLS

The experimental design for testing peripheral lymphocyte killing has been similar in different laboratories:
Fibroblasts or Hela cells carrying a persistent measles
virus infection are cocultivated with an excess of freshly isolated lymphocytes for 12 - 18 hrs without addition
of specific antibodies. The degree of cytotoxic activity
is determined by standard Cr-51 release. We would like
to summarize our experimental evidence using persistently measles virus-infected human embryonic lung fibroblasts (10).

As demonstrated in Figure 1, cytolytic activity is
low at 4 hrs and increases in a linear fashion over the
next 12 hrs. This suggests that a population of killer
cells might be generated by prolonged contact with
measles virus-infected cells. The time course of killer
cell activation is almost identical for patients with
SSPE, measles sero-positive adults and measles non-
immune children. Cell separation experiments revealed
that cytotoxicity was dependent on lymphoid cells carrying receptors for the Fc part of IgG (FcR γ) (8,11). By
further cell separation the bulk of cytotoxicity was
found to reside in the non-T cell fraction, while a

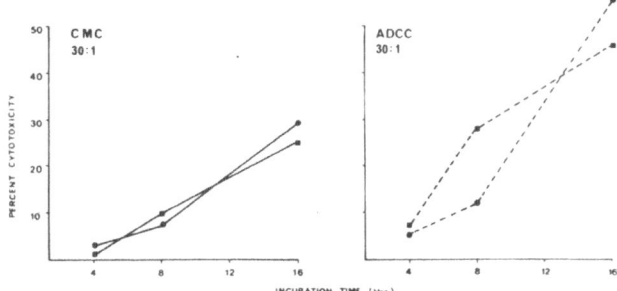

Figure 1. Generation of killer cells against persistently measles virus infected allogeneic target cells. A 30 x excess of either normal (⬤--⬤) or SSPE (▮--▮) peripheral blood lymphoid cells were incubated with Cr-51 labeled, persistently measles virus-infected human embryonic lung fibroblasts (10). CMC=cell mediated cytotoxicity in the presence of 10 % fetal calf serum. ADCC=antibody-dependent cellular cytotoxicity in the presence of 10 % fetal calf serum plus 2,5 % heat inactivated measles convalescence serum (HI titer 1:512). Percent cytotoxicity calculated according to (11).

smaller portion belonged to the T γ cell population (Figure 2). Tμ cells which comprise ca. 60 - 80 % of total T cells were found not to be cytotoxic.

It was originally speculated that the cytotoxicity observed might be due to K cells and a small amount of measles-specific antibody either locally produced in culture or otherwise passively transferred with lymphoid cells (11). It now seems that the cytotoxic effect in this system might actually be due to natural killer cells (NK cells). This conclusion is based on 3 lines of evidence: 1) Cytotoxicity is not restricted by the immune history of lymphocyte donors; 2) similar generation of cytotoxic potential can be achieved by cocultivation of peripheral lymphocytes with a number of tumor or other virus-infected cells (12, 13, 14); 3) the distribution of natural killer cells as

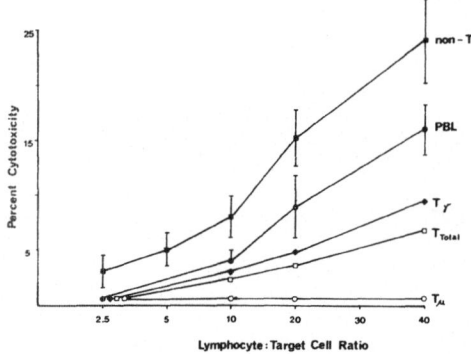

Figure 2. Cytotoxic activity of lymphocyte subpopulations against persistently measles virus-infected human embryonic lung fibroblasts. SSPE peripheral blood lymphocytes were separated into T and non-T cells by rosetting with neuraminidase-treated sheep red blood cells (25). T cells were further fractionated into Tγ and Tμ cells as described by MORETTA et al. (26). Lymphocyte subpopulations were tested for cytotoxic activity for 18 hrs in the absence of anti-measles antibodies. Means of triplicates $\pm$ 1 SD or means of duplicate determinations.

---

defined by cytotoxicity against the K 562 cell is also restricted to non-T and Tγ cells (15).

It should be stressed that up to now nothing is known about the biological role of natural killer cells in vivo. This type of effector cell must be clearly differentiated from virus-specific cytotoxic T lymphocytes.

4. CTL RESPONSE LIMITED TO THE ACUTE PHASE OF MEASLES

In mice, cytotoxic T lymphocytes are functionally restricted by the H-2 complex. Lysis will only occur if effector and target cells share either H-2D or H-2K gene products (16). It might be argued that in the experiments cited above with human lymphocytes, virus-specific CTL had remained undetectable because tests were always

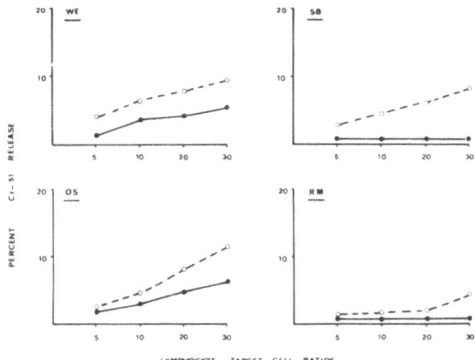

Figure 3. Peripheral lymphocyte killing in SSPE is not
restricted by HLA. Peripheral blood lymphoid cells from
4 patients with SSPE were tested on either autologous
(●———●) or allogeneic incompatible (○- - -○) measles
virus-infected blast cells in 4 hr Cr-51 release assays.
Autologous and allogeneic target cells contained almost
the same percentages of virus-infected cells as revealed
by the fluorescent antibody technique.

---

done against allogeneic target cells. The system was therefore modified to allow cytotoxic interactions to take place
under histocompatible conditions by using measles virus-
infected phytohemagglutinin-induced lymphoblasts with known
HLA determinants as target cells in 4 hr Cr-51 release
assays. There are several advantages in working with PHA-
stimulated lymphoblasts: once activated, lymphocytes are
permissive to measles virus infection regardless of the
immune history of the donor, and these cells are a poor
target for natural killer cells which ensures a lower background of Cr-51 release.

Freshly isolated lymphocytes from patients with SSPE
were definitely not restricted by HLA when tested on autologous or allogeneic histoincompatible measles virus-infected PHA blasts (Figure 3). Of particular interest is the
small but consistently higher cytotoxicity on allogeneic
incompatible   target cells by SSPE lymphocytes. This phenomenon can also be observed with effector cells from

136

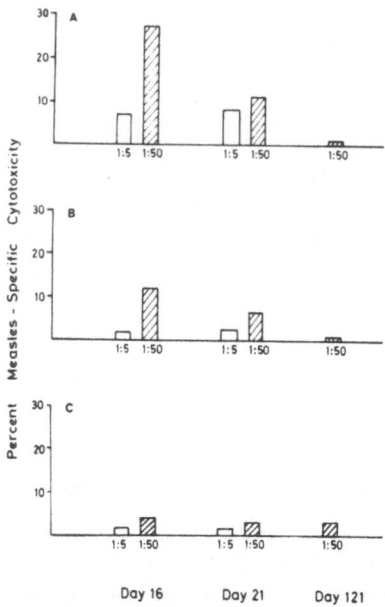

Figure 4. Kinetics and HLA dependency of killer cells in acute measles. Peripheral blood lymphoid cells were isolated from a child (HLA phenotype A1, AW24, B8, BW 44) at day 16, day 21 and day 121 after exposure to measles virus. Onset of measles rash was on day 14. Cells were cryopreserved in the presence of 11 % DMSO and 20 % FCS and stored in liquid nitrogen (17). After thawing, these cells were simultaneously tested in 4 hr Cr-51 release assays against uninfected and measles virus-infected PHA blasts bearing different HLA determinants. A: Autologous target cells prepared from day 121 lymphocytes. B: Semi-allogeneic target cells, HLA type A2, AW31, B8, BW42. C: Allogeneic incompatible target cells, HLA type A2, A26, B15, BW22. Tests were run at effector-target cell ratios of 1:5 and 1:50. Background killing of uninfected PHA blasts was in the range of 1 - 2 %.

normals. The reason for this "allogeneic preference" by
natural killer cells is not known.

Markedly different results were obtained with peripheral blood lymphocytes from patients with acute measles(17).
As demonstrated by one representative example (Figure 4)
cytotoxicity depends on 2 important variables: 1) the
stage of acute disease when effector lymphocytes are obtained and 2) the extent of histocompatibility between
effector and target cells. As shown in Figure 4A, cytolytic activity of lymphocytes obtained 3 days after measles
rash (day 16) is much higher than that of cells collected
5 days later (day 21). A final blood sample donated more
than 3 months later (day 121) was devoid of killer cells.
These data which could be confirmed in other patients
suggest a rapid decline of cytotoxic cells during early
convalescence. The speed with which cytotoxicity declines
may differ from one patient to another and sampling time
seems crucial. This might be the reason why others (18)
working in a similar system could not detect any cytolytic
activity by cells from acute measles though testing was
done under HLA defined conditions.

By fluorescent antibody technique and antibody and
complement mediated lysis all target cells were found to
contain almost the same percentages of cells bearing sur-
face-bound viral antigens, however, not all virus-infected
blast cells were susceptible to lysis. Cytolytic activity
was clearly dependent on the degree of histocompatibility
between effector and target cells in a hierarchal pattern:
Highest cytolysis was found on autologous target cells
(Figure 4A), while lysis was minimal on allogeneic incompatible target cells (Figure 4C). Sharing of a single
antigen between effector and target cells (e.g. HLA B8
in Figure 4B) elicited an intermediate reaction. It seems
that HLA dependent killing in measles is associated both
with HLA A and B determinants. Effective cytolysis was
observed when effector and target cells shared HLA A2, A3,
AW24, A28, B5, B8, B12, BW35.

HLA dependent killer cells from patients with acute

138

measles cannot be adsorbed onto plastic-bound IgG immune
complexes (Figure 5). These cells therefore belong to the
FcR γ negative pool of lymphocytes. Similar properties
have been reported for virus-specific and alloreactive
cytotoxic T cells in mouse and man (19,20,21). It should
be noted that killing by SSPE peripheral lymphocytes of
autologous virus-infected target cells is totally abrogat-
ed by a similar adsorption procedure (Figure 5, lower
half).

In summary, killer cells in acute measles have 3 characteristic features: 1) Dependence on the acute stage of
the disease, 2) functional restriction by the HLA system,
and 3) absence of Fc γ receptors. We think that these
cells represent genuine virus-specific T lymphocytes with
strikingly similar properties to CTL in other species (16).

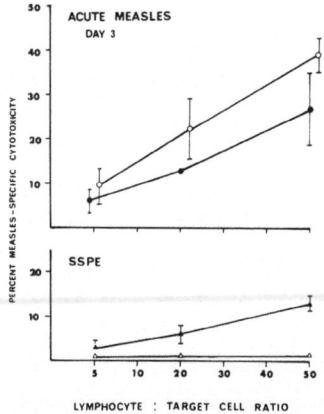

Figure 5. HLA dependent killer cells in acute measles belong to the FcR γ negative cell population. Lymphocytes
were tested either unfractionated (●——● , ▲——▲) or
after removal of FcR γ bearing lymphoid cells by adsorption to solid IgG immune complexes (○——○ , △——△). 4 hr
Cr-51 release assays were run on autologous measles virus-
infected and uninfected PHA blasts. Background killing of
uninfected target cells was in the range of 1 - 2 %.

5. GUIDELINES FOR FUTURE RESEARCH

With measles-specific cytotoxic T lymphocytes confined to acute measles the question is then raised on
how to pursue studies on cell-mediated immunity in
SSPE. Again, experimental clues can be derived from
animal studies. Experiments in mice have shown that
although cytotoxic T cells disappear from lymphoid
tissues within 20 days after virus infection, memory
is still retained. These memory T cells can be reacti-
vated by appropriate antigenic stimulation in vitro
into secondary populations of cytotoxic T cells (22).

Thus efforts should also be made in man to reinduce
measles-specific cytotoxic T lymphocytes in vitro. As
has been demonstrated for influenza virus in man, virus-
specific memory T cells can be found within the pool
of recirculating lymphocytes (23,24). Such indirect
cytotoxicity tests with measles convalescent and SSPE
lymphocytes would be the crucial challenge to BURNET's
hypothesis (1). If BURNET's assumption of a specific T
cell deficiency is correct, memory for T cell cytotoxicity should not exist in SSPE.

ACKNOWLEDGEMENTS

These studies were supported by grant Kr 376/8 from
the Deutsche Forschungsgemeinschaft. H. Pabst is a
recipient of the Thyssen Foundation stipend. We wish
to thank Lore Kress for excellent technical assistance
and Helga Schneider for typing the manuscript.

REFERENCES

1. Burnet, FM: Measles as an index of immunological function. Lancet ii: 610-613, 1968.
2. Agnarsdottir, G: Subacute sclerosing panencephalitis. Recent Adv Clin Virol 1: 21-49, 1977.
3. Simpson, E, PC Beverley: T cell subpopulations. In: Progress in Immunology, III, Amsterdam, North-Holland, 1977, p 206-214.
4. Burns, WH, LC Billups, AL Notkins: Thymus dependence of viral antigens. Nature 256: 654-655, 1975.
5. Valdimarsson, H, G Agnarsdottir, PJ Lachmann: Cellular immunity in subacute sclerosing panencephalitis. Proc Roy Soc Med 67: 1125-1129, 1974.
6. Kreth, HW, YM Kaeckell, V ter Meulen: Cellular immunity in SSPE patients. Med Microbiol Immunol 160: 191-199, 1974.
7. Steele, WR, DA Fuccillo, SA Hensen, MM Vincent, JA Bellanti: Specific inhibitory factors of cellular immunity in children with subacute panencephalitis. J Pediat 88: 56-62, 1976.
8. Perrin, LM, A Tishon, MBA Oldstone: Immunologic injury in measles virus infection. III. Presence and characterization of human cytotoxic lymphocytes. J Immunol 118: 282-290, 1977.
9. Ewan, PW, PJ Lachmann: Demonstration of T cell and K cell cytotoxicity against measles-infected cells in normal subjects, multiple sclerosis and subacute sclerosing panencephalitis. Clin Exp Immunol 30: 22-31, 1977.
10. Norrby, E: A carrier cell line of measles virus in Lu 106 cells. Arch ges Virusforsch 20: 215-224, 1967.
11. Kreth, HW, G Wiegand: Cell-mediated cytotoxicity against measles virus in SSPE. II. Analysis of cytotoxic effector cells. J Immunol 118: 296-301, 1977.
12. Morales, A, GD Bonnard, M Dean, JH Herbermann: In vitro augmentation of natural killer (NK) and antibody-dependent killer (K) cell activities by co-cultivation with mumps virus. (Abstract) Fed Proc 36: 1325, 1977.
13. Trinchieri, G, D Santoli: Anti-viral activity induced by lymphocytes with tumor-derived or virus-transformed cells. Enhancement of human natural killer cell activity by interferon and antagonistic inhibition of susceptibility of target cells to lysis. J exp Med 147: 1314-1333.

14. Santoli, D, G Trinchieri, FS Lief: Cell-mediated
    cytotoxicity against virus-infected target cells in
    humans. I. Characterization of the effector lympho-
    cyte. J Immunol 121: 526-531, 1978.
15. Kall, MA, HS Koren: Heterogeneity of natural killer
    cell populations. Cell Immunol 40: 58-68, 1978.
16. Doherty, PC, RV Blanden, RM Zinkernagel: Specificity
    of virus-immune effector T cells for H-2K or H-2D
    compatible interactions. Implication for H2-antigen
    diversity. Transpl Rev 29: 89-124, 1975.
17. Kreth, HW, V ter Meulen, G Eckert: Demonstration of
    HLA restricted killer cells in patients with acute
    measles. Med Microbiol Immunol 165: 203-214, 1979.
18. Galama, JMD, CT Lucas, A Vos: Lymphocyte-mediated
    cytotoxicity to cells infected with measles virus.
    I. Use of in vitro infected leukocytes as autologous
    target cells; absence of virus-specific cytotoxic
    lymphocytes. Cell Immunol 38: 365-377, 1978.
19. Kedar, E, M Ortiz de Landazuri, B Bonavida: Cellular
    immunoadsorbents: A simplified technique for sepa-
    ration of lymphoid cell populations. J Immunol 112:
    1231-1243, 1974.
20. Ertl, M, M Koszinowski: Cell-mediated cytotoxicity
    against Sendai-virus-infected cells. Z Immun Forsch
    152: 128-140, 1975.
21. Eijsvoogel, VP, TA Schellekens, MJGJ Du Bois, WP Zeijle-
    maker: Human cytotoxic lymphocytes after alloimmuni-
    zation in vitro. Transplant Rev 29: 125-145, 1975.
22. Dunlop, MBC, RV Blanden: Secondary cytotoxic cell re-
    sponse to lymphocytic choriomeningitis. I. Kinetics
    of induction in vitro and yields of effector cells.
    Immunol 31: 171-180, 1976.
23. McMichael, AJ, B Askonas: Influenza virus-specific
    cytotoxic T cells in man: induction and properties
    of the cytotoxic cell. Europ J Immunol 8: 711-715,
    1978.
24. Biddison, WE, S Shaw, DL Nelson: Virus specificity of
    human influenza virus-immune cytotoxic T cells.
    J Immunol 122: 660-664, 1979.
25. Gmelig-Meyling, F, RE Ballieux: Simplified procedure
    for the separation of human T and non-T cells.
    Vox Sang 33: 5-8, 1977.
26. Moretta, L, M Ferrarini, MC Mingari, A Moretta,
    SR Webb: Subpopulations of human T cells identified
    by receptors for immunoglobulins and mitogen re-
    sponsiveness. J Immunol 117: 2171-2174, 1976.

DISCUSSIONS OF PAPERS BY P. J. LACHMANN ET AL. AND
H. W. KRETH

This began with a query as to whether the blood-brain barrier was
altered in SSPE and, if so, whether this was a selective lesion,as in
malignant meningitis or the nephrotic syndrome, or not. The question
was also put as to whether immunoglobulins could be synthesised in
the central nervous system. In MS and SSPE, this was almost certainly
so, as the IgG excess was calculable and had to come from within the
CNS. The evidence suggested that oligoclonal immunoglobulin was
produced in the CSF; bands were found in CSF protein which were not
found in serum but bands found in serum were usually also present in
the CSF. It was possible that some of the immunoglobulin was formed
inside the CNS, and some outside. The amount of antigen which was
synthesised in LCM and measles was considered and it was suggested
that probably only a small amount was synthesised outside the CNS
in SSPE. There was no idea how much antigen was required to stimulate
hyper-immune reactions like those that occur in LCM and SSPE.

The question of HLA subtypes and their relationship to SSPE was
raised. A prospective study of measles in children might be productive and there are some data on survivors of the congenital rubella
syndrome who have a somewhat similar syndrome to SSPE. Professor ter
Meulen mentioned one child in his experience who, one year before the
onset of SSPE, was tested for an entirely different reason and was
found to have a very high level of measles antibody. Dr Weiss
commented that siblings of children with SSPE were less often affected
than might be expected; Dr Kreth, in reply, wondered whether HLA subtypes were involved in the host response. Following further discussion
on the nature of the immune reaction in SSPE, possible modification
of the disease process by immunological stimulation was suggested by
Professor ter Meulen. In reply, Dr Lachmann mentioned that
Dr Valdemarsson had used transfer factor in some patients; this
might have slowed down the disease process but had no real effect.

A question was asked concerning the possible blocking of the T
cell response by infected cells being coated with antibody; this did
not occur.

Dr Norrby summed up the pathogenesis of SSPE as he saw it.
Following measles infection in early childhood (under the age of two

years), a persistent infection in a few target cells in the CNS arose
in some individuals, usually boys, where there was an immature immune
system which could not eliminate such an infection. At the cellular
level, viral antigen was not exposed at the cell surface and,
associated with deficient virus matrix protein synthesis, antigen
accumulated within the cell. By cell to cell contact, over a period
of five to seven years, infection of more cells occurred and a larger
amount of antigen accumulated. This was released, leading to hyper-
immunisation, break-down of CNS tissue and the precipitation of the
clinical features of SSPE. In the persistent infection described,
changes in the matrix protein might occur, possibly affecting glycoprotein synthesis and the expression of glycoprotein at the cell
surface.

THE BIOLOGY OF RNA TUMOUR VIRUSES

R. A. WEISS

1. INTRODUCTION

RNA tumour viruses, or oncoviruses as they are now called,
belong to the family Retroviridae, comprising the retroviruses which also include foamy virus (*Spumavirinae*) and
Maedi/Visna virus (*Lentivirinae*). Oncoviruses are divided
according to a morphological classification into Type B, C
and D viruses (1). Type C viruses have been isolated from
or identified in numerous vertebrate hosts ranging from
fish to mammals, and also from mosquitos. Oncoviruses are
known to cause a variety of neoplasms in their natural host
species. Thus the lymphomatous leukoses of cattle, cats,
mice and chickens are typically caused by Type C oncoviruses, whereas Type B and D oncoviruses are associated with
mammary carcinomas. Rare, acute neoplasms, such as sarcomas,
and erythroid and myeloid leukaemias, are also recognised to
result from retrovirus infection, as well as non-malignant
diseases, such as osteopetrosis in chickens, anaemia in
cats, and possibly autoimmune and paralytic diseases in
mice. The problem of identifying retroviruses with neoplastic potential in humans remains equivocal, although tanta-
lising items of evidence continue to be thrown up, as exemplified by Thiry's contribution to this volume. While
much of the impetus and funding for research in viral on-
cology is motivated by the search for human retroviruses,
retrovirus research today is proving most useful in providing conceptual models of oncogenesis and experimental systems for probing the molecular and cell biology of neoplasia.

2. TRANSMISSION OF RETROVIRUSES

Although the genetic information of retrovirus particles is
contained in RNA molecules, upon infection of the host cell
this information is transcribed by the viral enzyme, RNA-
directed DNA polymerase (reverse transcriptase) into a
double-stranded DNA provirus (2). This provirus, like the

genomes of DNA tumour viruses, becomes inserted into host
chromosomal DNA, so that the "integrated" viral genes
become adopted by the host as extra genetic information.
Integration is probably not the oncogenic event itself, although the insertion of new DNA sequences at inappropriate
sites could very conceivably cause disruption of cellular
regulatory mechanisms. Nevertheless, integration is the
means by which viral genes may be heritably transmitted to
daughter cells. Furthermore, during the evolution of
retrovirus-host relations, retrovirus genomes have on
occasion become integrated into cells of the host germ line,
with the result that the viral genes are now inherited from
one generation to the next as host Mendelian factors. Such
stable, inherited viral genomes (called endogenous viruses),
with the exception of certain inbred strains of mice, are
not known to be oncogenic, but may give rise to oncogenic
agents on reactivation to viral form, and may become re-
combined with other DNA sequences to form new genetic
elements that are potentially oncogenic.

Thus retroviruses can persist by masquerading as host
genetic information. These endogenous viral genes may be
unexpressed for many host generations, or some viral antigens may be synthesised in certain types of host cell. On
occasion, complete virus may be activated, either spontaneously, or by treatment of the host cell with ionizing
radiation or chemical carcinogens and mutagens. The reactivation phenomenon led to an hypothesis that all cases of
oncogenesis by diverse agents might be accounted for by
activation of endogenous viruses (3). This now seems unlikely, and the most efficient virus-inducing agents, such
as halogenated pyrimidines, have little carcinogenetic
potential. The latency and inheritance of retrovirus
genomes considerably complicates any analysis of epidemiology, not least because some of the newly activated viruses
frequently cannot reinfect cells of the species in which
they are inherited, but may be infectious for foreign
species, a phenomenon called xenotropism (4).

Leukaemogenic retroviruses, with the important exception of murine leukaemia viruses, are typically transmitted
as infectious agents. Thus leukosis is a contagious disease
in cats and cattle which is spread horizontally by close
contact with infected individuals. Horizontal infection of
chickens results most frequently in effective viral immunity, but 'vertical' infection of eggs leads to immunological
tolerance and perhaps as a consequence of a persistently
high viral load, such congenital infection typically causes
a bursal lymphoid leukaemia. Congenital retrovirus infection
also occurs in several mammalian species via either the
placenta or the milk, and activation of endogenous virus in
mouse embryos can even lead to a 'reverse vertical' infection of a non-viraemic mother.

The evolutionary origin of retroviruses is often
obscure. The exogenous retrovirus causing bovine leukaemia
appeared as a new enzootic agent in Danish cattle some
years ago and might have been transmitted from another
species. Endogenous viral genomes have frequently been
present in the germ line of their hosts for many million
years, showing divergence of genetic sequences alongside
host gene divergence during evolution. On the other hand,
some endogenous viruses have obviously been acquired by
exogenous infection emanating from unrelated host species in
recent evolutionary times. Thus the endogenous virus of
cats has been acquired from ancestors of modern baboons (5).
The endogenous virus of chickens comes from an unknown
source; it is present in the feral, undomesticated form, the
Red Junglefowl, but not in other Junglefowl species (6).
Exogenous chicken leukosis viruses are very closely related
genetically to the endogenous chicken virus, but differ in
host range and leukaemogenicity; presumably they are derived
from the endogenous virus. On the other hand, the exogenous
feline leukaemia virus strains are quite unrelated to the
endogenous feline virus, and may be derived from endogenous
rodent viruses. Clearly there has been much hopping in and
out of host genomes of different species in the evolution

and spread of retroviruses, and it is possible that these
viruses act as vectors of parasexual genetic exchange
between phylogenetically distant host species. Mendelian
inheritance of retrovirus genomes is the ultimate form of
persistant infection, so that viral genes may become subject to natural selection acting on the host species if
they are enrolled to perform host functions. For instance,
endogenous retroviral glycoproteins in mice may play a role
in the maturation of lymphocytes (7) and in the function of
certain secretions (8).

3. RETROVIRUS GENES
The proteins of retroviruses are, of course, antigenic, and
the preparation of specific antisera for radioimmunoassays
and other immunological techniques has been of great use
for studying viral gene expression and virus relationships.
The other major analytical tool has been nucleic acid
hybridisation. With the preparation of specific radioact-
ive probes, the presence and expression of retroviruses and
of single retrovirus genes can be accurately monitored, and
the evolutionary relationships between retroviruses in
different host species can be assessed.

   Non-defective retroviruses have a simple unit genome
comprising three well-defined genes coding for virion proteins (2). The *gag* gene encodes a large, precursor polypeptide which becomes proteolytically cleaved to generate
the internal or core antigens of the virion. These proteins are named according to their estimated molecular
weight, e.g. murine p30 denotes the major core protein of
30,000 daltons of murine leukaemia virus (MuLV). Precursor
polypeptides are similarly labelled pr65, or pr90, etc. The
env gene encodes the proteins located in the envelope of the
virion which is derived by budding from the plasma membrane
of the host cell; thus murine gp70 denotes the glycosylated
envelope protein of MuLV of approximately 70,000 daltons.
The *pol* gene encodes the polymerase (reverse transcriptase).

The three genes are ordered in the genomic RNA molecule
in the sequence 5'-*gag-pol-env*-3'; apart from some nucleotide seqeunces at the 5' end of the molecule, they appear
to be translated from separately transcribed mRNA species.

4. ONCOGENESIS

Retroviruses can be roughly divided into three groups on
the basis of oncogenicity, 'non-transforming', weakly transforming' and 'strongly transforming' viruses. Those that
cause acute neoplasms with short latent periods between infection and the appearance of the tumour are called strongly transforming viruses. In most cases these viruses will
also transform appropriate target cells in culture. Weakly
transforming viruses cause tumours only after long latent
periods, i.e., months rather than days in mice and chickens,
the most closely studied host species, and *in vitro* transformation systems have not to date been devised for these
viruses. Some endogenous C-type retroviruses such as
those of cats and chickens, may be regarded as non-transforming, but this classification may have to be modified
when they are studied in more detail.

4.1. *Strongly transforming viruses*

These viruses rarely occur in nature, but their recognition,
isolation and experimental use has led to major advances in
our understanding of viral oncogenesis. The best known and
most venerable example of a strongly transforming virus is
the Rous sarcoma virus (RSV) of chickens (9), others are
the avian myeloblastosis and erythroblastosis viruses, and
the murine Friend erythroleukaemia, Abelson lymphoma and
Moloney, Kirsten and Harvey sarcoma viruses. RSV has a
gene, designated *src* for sarcoma induction, in addition to
the three genes essential for viral replication. Studies
of deletion mutants and temperature-sensitive mutants have
shown that the *src* gene is essential for fibroblast transformation and for sarcomagenesis, but is not required for
viral replication. Recent data indicate that the *src* gene
product is a cytoplasmic protein of 60,000 daltons that
possesses protein kinase activity (10). Precisely how this

protein causes cell transformation and what are the crucial targets in the cell for phosphorylation remains to be determined. Nevertheless it is a remarkable advance in experimental oncogenesis that an enzyme has been identified with an oncogene.

In most strains of RSV, the *src* gene is carried as an extra gene to the viral genes, in the order 5'-*gag-pol-env-src*-3'. All other well studied strongly transforming viruses are defective for replication, that is, new genetic information specifying neoplastic transformation (*onc* genes) appears to be inserted in the viral genome in place of essential genes for replication. A part of the *gag* gene and the 3' end of the viral genome are usually maintained, giving a typical structure 5'-*ga-onc*-3'. The infectivity of such defective viruses relies on the presence of replication-competent 'helper' viruses, and the disease spectrum caused by such viruses depends on the properties of this complex virus population. In several strains of defective virus, the polypeptide coded by the supposed *onc* region of the viral genome actually starts in what remains of the *gag* gene. Since the polypeptide therefore bears some *gag* antigens, it can be identified by immunoprecipitation with anti-*gag* anti-sera from lysates of transformed cells. Such *gag-onc* 'polyproteins' have been detected in cells transformed by avian myelocytoma and erythroleukaemia viruses, murine Abelson leukaemia cells, and cells transformed by feline sarcoma virus. These proteins are not related to the *src* protein of RSV or to each other, and each may have individual functions resulting in neoplastic transformation. This would account for the high degree of specificity of the target cell for transformation, as each virus causes a specific type of cancer or leukaemia.

Oncogenes appear to originate from the host, as genetic elements related by molecular hybridisation to viral oncogenes are found in the host genome, though not linked to endogenous viral elements (13). Possibly the natural host sequences code for normal proteins important in the

function or differentiation of particular cell types (15).
When they are picked up and modified by viral genomes, and
reinserted into appropriate target cells, they may cause
disruption to regulatory cell functions, blocking or even
reversing the normal pathway of differentiation. Further
analysis of the oncogenes of strongly transforming viruses
should illuminate much about differentiation and neoplasia.

4.2. *Weakly transforming viruses*

These viruses do not appear to carry oncogenes distinguish-
able from the three viral genes, *gag*, *pol* and *env*. Commonly
occurring weakly transforming viruses are the murine mammary
carcinoma virus, murine thymic lymphoma viruses, avian
bursal leukosis viruses, and the leukaemia viruses of cats
and cattle. In contrast to the strongly transforming viruses, the tumours they cause appear after long latent periods
and only very few of the cells that become infected subsequently give rise to tumours. The tumours are probably
clonal in origin, whereas with strongly transforming
viruses such as RSV the tumours grow as quickly by infection and transformation of new target cells, as by
mitosis of the originally transformed cell. There is growing evidence for the murine viruses which induce thymic
lymphomas that genetic recombination involving the *env* gene
takes place, often between xenotropic and mouse-tropic endogenous viruses, giving rise to new virus variants which
may interact with and transform different cell types than
those transformed by the parental viruses (15). The re-
combinant *env*-coded glycoproteins may play a dual role,
both by allowing the virus to recognise and infect specific
target cells bearing appropriate receptors for the glycoproteins and by acting as a perpetual mitogenic stimulus to
such cells.

In summary, oncoviruses persist so intimately with
their hosts, that genetic exchange in both directions has
taken place. On the one hand, viral genomes have become
adopted and perhaps exploited as new host genes; on the
other hand, host genetic sequences have been incorporated
into viral genomes, which in some cases become highly oncogenic.

References

1. Tooze, J: The molecular biology of tumour viruses, Cold Spring Harbor Laboratory 1973.

2. Baltimore, D: Tumor Viruses, Cold Spring Harbor Symp. Quant. Biol. 39:1187-1200, 1974.

3. Todaro, GD: and RJ Huebner: The viral oncogene hypothesis: new evidence. Proc. Nat. Acad. Sci. U.S.A. 69:1009-1015, 1972.

4. Weiss, RA: Receptors for RNA tumor viruses. In: Cell membrane receptors for viruses, antigens and antibodies, polypeptide hormones, and small molecules, Beers, RF, EG Bassett (eds), Raven Press, New York, 1976, p237-251.

5. Todaro, GJ: RNA tumour virus genes and transforming genes: patterns of transmission. Br. J. Cancer 37: 139-158, 1978.

6. Frisby, DP, RA Weiss, M Roussel, D Stehelin: The distribution of endogenous chicken retrovirus sequences in the DNA of galliform birds does not coincide with avian phylogenetic relationships. Cell, in press.

7. Moroni, C, G Schumann: Are endogenous C-type viruses involved in the immune system? Nature 269:600-601, 1977.

8. Elder, JH, FC Jensen, ML Bryant, RA Lerner: Polymorphism of the major envelope glycoprotein (gp70) of murine C-type viruses: Virion associated and differentiation antigens encoded by a multi-gene family. Nature 267:23-28, 1977.

9. Rous, P: A sarcoma of the fowl transmissible by an agent from the tumor cells. J. Exp. Med. 13:397-402, 1911.

10. Collett, MS, RL Erikson: Protein kinase activity associated with the avian sarcoma virus src gene product. Proc. Nat. Acad. Sci. U.S.A. 75:2021-2024, 1978.

11. Bister, K, MJ Hayman, PK Vogt: The defectiveness of avian myelocytomatosis virus MC29: isolation of long term non-producer cultures and analysis of virus specific polypeptide synthesis. Virology 82:431-448, 1977.

12. Witte, ON, N Rosenberg, M Paskind, A Shields, D Baltimore: Identification of an Abelson murine leukemia virus-encoded protein present in transformed fibroblast and lymphoid cells. Proc. Nat. Acad. Sci. U.S.A. 75:2488-2492, 1978.

13. Stehelin, D, HE Varmus, JM Bishop, PK Vogt: DNA related to the transforming gene(s) of avian sarcoma viruses is present in normal DNA. Nature 260: 170-173, 1976.

14. Risser, R, E Stockert, LJ Old: Abelson virus: a viral tumor antigen that is also a differentiation antigen of BALB/c mice. Proc. Nat. Acad. Sci. U.S.A. 75: 3918-3922, 1978.

15. Elder, JH, JW Gautsch, FC Jensen, RA Lerner, JW Hartley, WP Rowe: Biochemical evidence that MCF murine leukemia viruses are envelope (env) gene recombinants. Proc. Nat. Acad. Sci. U.S.A. 74: 4676-4680, 1977.

EVIDENCE FOR THE PRESENCE OF RETROVIRUS MARKERS IN MAN

L. THIRY

Three simian viruses which may be related to human retrovirus strains.

Virologists in search of retroviruses in man must bear in mind that these viruses may be either endogenous with xenotropic properties, or exogenous with ecotropic characteristics. A simian virus of the first type is the Baboon endogenous virus (BeV), the complete genome of which is present in the DNA of all tissues of monkeys taxonomically related to baboons (1). However, the virion is expressed only during pregnancy and then only in the placenta (2). Even in this tissue, the cells are probably not highly permissive for BeV replication, for continuous production of the virus has been obtained only by cocultivation of baboon placentas with cells from other animal species, among which the human rhabdomyosarcoma cell line A204. Using cell hybrids formed from the fusion of mouse cells with a human cell line, (VA) infected with BeV, it was possible to demonstrate that the BeV genome had been integrated in human chromosome 6 (3). However, ^{125}I labelled RNA from BeV hybridized poorly with the DNA of normal human organs (4). Two agents will be cited as examples of exogenous viruses with ecotropic properties, Mason-Pfizer virus (MPMV) and Simian Sarcoma virus (SiSV). The characteristics of MPMV have been reviewed recently (5). This virus was originally isolated from a breast carcinoma of a rhesus monkey. Although it can also be found in placentas and lactating glands of pregnant monkeys it is transmitted horizontally, not vertically. The genome of MPMV is not present in the DNA of normal tissues in the natural hosts. It multiplies freely in cells from monkeys and humans, but not in cells from the many other mammals tested thus far. Although MPMV possesses the capacity to transform monkey and human cells in vitro, attempts to induce tumours in various strains of monkeys have been unsuccessful. In contrast, the Simian Sarcoma virus, first isolated from a fibrosarcoma of a woolly monkey (6) can induce tumours in newborn marmosets (7). This virus is replication-defective, but in the presence of a helper virus (Simian Sarcoma-associated virus) it multiplies to high titers in human and other mammalian cells.

Retrovirus footprints in human cancerous and normal tissues.

The complicated results obtained with leukaemic cells have been reviewed recently (8). Attempts to isolate infectious virions must take into account the possibility that tumour cells may contain defective or xenotropic viruses. The multiplication of defective virus can only be obtained in the laboratory by deliberate addition of helper viruses. To detect xenotropic viruses, a wide range of animal cells must be tested empirically. Alternately, one may attempt to produce pseudotypes by cultivating the tumour cells in the presence of a virus which replicates in human cells. The latter virus can provide adequate envelopes for the putative viral genomes originating from the tumour tissues. However, these laboratory manipulations may introduce other complications. The added laboratory strains may yield recombinants with the patient's viral genes. Sometimes, introduction of viral genes in the experimental protocol may be accidental. This might have occurred when HL-23 virus was isolated after cultivation of cells from a case of acute myeloid leukemia in medium conditioned by a unique human embryo cell strain (9). As discussed below, the possibility that a virus was present in the conditioning medium was made more likely by accumulating evidence that retro viruses are preferentially induced during embryogenesis and differentiation. Actually, the HL-23 isolate was found to contain two viruses, one closely related to SiSV and the other to BeV (10). The obvious tactic was to look for markers of BeV and SiSV in the tissues of leukaemic patients. DNA from several patients with leukaemia hybridized 70 % of the hybridizable RNA from BeV, and Tm of the hybrids was high, while only 23 % of the BeV RNA hybridized to DNA of normal tissues, with a lower Tm (11). To our knowledge, the latter high figure for BeV nucleotide sequences in normal human tissues has not been confirmed by others. SiSV information was not found in the form of proviral DNA in leukaemic or other human tissues. However , cytoplasmic particles hybridizable to SiSV probes were detected in leukaemic cells from 2 patients, but such RNA sequences were also found in normal lymphocytes stimulated in vitro with PHA. The presence of retroviruses can also be demonstrated with the use of the XC or KC plaque assays. SiSV viruses form syncytia on the XC cells, while MPMV and BeV produce syncytia on the KC cells. Bone marrow cells from a child with acute lymphoblastic leukaemia formed plaques on XC cells after stimulation with PHA (8, 12). Negative results were obtained with similarly treated cells from 7 adults with other types of leukaemia and 4 normal adult donors. The samples were not seeded on KC cells, so there is no evidence for or against the presence of a viral component biologically related to MPMV or to BeV.

Several authors have attempted to demonstrate the presence of viral genes by cocultivating the tumour cells with putative permissive cells. In these experiments, attempts are made to insure close initial contact between the two types. We suggest that efficiency of the method might be increased, if the cells were fused with polyethylene glycol. To our knowledge, no such attempt has been reported in the literature. If one looks for infectious virions in the supernatants of tumour cell cultures, these supernatants should be seeded on putative permissive cells pretreated with polybrene to increase the chances of virus penetration. If whole genomes of retroviruses were present in cancer cells, but without production of enveloped virions, infectious DNA should be demonstrable by transfection experiments. By this technique, no positive results were obtained from the leukemic cells of 12 patients, nor did the extracted DNA recombine with or rescue endogenous human virus or BeV (13). The possibility must be considered that only some of the viral genes are integrated into the host genome. It may be possible to detect RNA transcripts of such genes with the newly developed in situ hybridization technique.

Attempts to detect viral coded proteins have led to conflicting results. Extracts from peripheral blood leukocytes of 5 patients with acute leukemia competed with labelled p30 related to SiSV in a radioimmunoassay (14). In similar assays, a p30 related to those of known retroviruses was detected in both normal and malignant human tissues (15). However, proteolytic enzymes present in tissue extracts can break down the labelled polypeptides. The results of such experiments would then mimic the displacement of the labelled polypeptides by viral polypeptides thought to be present in the cell extracts. Nonspecific inhibitors which interfered with radioimmunoassays were found in several human tissue specimens (16).

Antigens related to SiSV (17) or to BeV (18) were detected in cultures of sarcoma cells treated with specific antisera in an indirect immunofluorescence test. Finally, an indirect demonstration of the presence of a retrovirus antigen on the surface of leukocytes from patients with chronic myelogenous leukaemia was obtained by demonstrating that immunoglobulins eluted from the leukaemia cells specifically neutralized the reverse transcriptase of feline leukaemia virus (19).

In regard to the problem of human breast carcinoma, we feel that it is premature to discuss the evidence for markers of mouse mammary tumour virus and/or of MMPV in this type of tumour.

Antibodies to retrovirus antigens in healthy individuals.

Here again we are confronted with apparently discrepant results, even though similar radioimmunoprecipitation procedures were used. By this method, antibodies to SiSV virions were found in most human sera, while the results were negative with purified gp 70 of the same virus (20). Part but not all of the reactivity with SiSV was removed after adsorption of the human sera with calf serum. This fits with the findings that serum components from culture medium become associated with budding retrovirus particles, and suggests that some humans have antibodies directed to calf proteins. Negative results with gp70 may be related to fragility of the molecule, and to the possibility that the chloramine T treatment during radioiodination procedure denatured the reactive determinant. Positive reactions with SiSV virions were confirmed by another group (21), who demonstrated antibodies to BeV particles in about half the adult sera tested. Lack of reactivity of human sera with internal viral proteins was apparent from data showing that human sera did not react with p30 of a virus related to SiSV (22) nor with p25 of MPMV (23).

A role of retroviruses in human reproduction?

a. Reverse transcriptase in spermatozoids.

A DNA polymerase with the properties of reverse transcriptase was isolated from the nuclei of human spermatozoids (25), and showed an immunological relationship to

BeV reverse transcriptase. Immunoglobulins from partners
of infertile marriages inhibited both sperm and BeV reverse
transcriptase. The authors postulated that the sperm polymerase plays a positive role during fertilization and/or
early embryogenesis.

b. Retroviruses during differentiation and embryogenesis

Soon after the discovery of reverse transcriptase, it
was suggested that this enzyme could play a physiological
role if present in normal cells. For instance, at some
stages of embryogenesis, a reverse transcriptase might use
ribosomal RNA as templates to reproduce DNA copies and
lead to gene amplification. This hypothesis has not been
substantiated by experimental results. However, the suggestion that retroviruses might play a part in normal cell
differentiation was strongly supported by the apparent lin-
kage between regulation of type C RNA virus production and
cell differentiation in mouse myeloid leukaemic cells (26).
Virus production was an early step in the induction of differentiation, indicating that it was a causative factor
rather than an effect. Also, the induction of murine B
lymphocyte differentiation into immunoglobulin secreting
cells was accompanied by budding and release of retrovirus
particles (27). However, attempts to demonstrate virus
induction in human lymphocytes have failed so far.

A diploid cell culture, HEL-12, was obtained from the
lungs of an 8-week-old human embryo and was frozen after
a primary growth cycle (28). Reinitiation of cell growth
from the frozen stock yielded virus expression after several subcultures and a lag period of 80-120 days. Expression
of viral antigens, reverse transcriptase and infectious
virus was cyclic. The virus grew on human and other mammalian cells, and was composed of a heterogeneous population with a SiSV-like and a BeV-related component.

We have studied 6 short-term cultures of 6-10 weeks old
human embryos, 2 of which were obtained by curettage
because of non-progressive pregnancies in women with repeated spontaneous abortions, and 4 were aspiration products
from women with social problems (unpublished results).
Semi-confluent cultures of heterogeneous cell populations
were obtained within 4-7 days. At that point, cells were
treated with 30 ug/ml of 5-iododeoxyuridine, in order to
increase the chances of retrovirus expression, and then
submitted to three tests. Culture media were concentrated
and assayed for reverse transcriptase activity in the presence of Mg^{++} or Mn^{++}. A portion of the cells was seeded
on KC and on XC cells to test for viral dependent syncytial cell formation. The remaining cells of the foetal
culture were labelled with ^{51}Cr and treated with anti-BeV
and anti-MPMV sera in the presence of complement. Indications of the presence of a BeV-related virus were obtained in 1 of the 2 cases of non-progressive pregnancy, and
3 of the 4 aspirates from the apparently healthy cases.
Positive evidence included : syncytium formation on KC
cells, ^{51}Cr release from the foetal cells exposed to anti-
BeV serum, and weak but significant Mn^{++} dependent reverse
transcriptase activity. Attempts to cultivate the virus(es)

in the human cell line A 204 are in progress.

c. Retrovirus markers in placentas.

Some of the described results tend to support the idea
that retroviruses play a part in early implantation of the
embryo and/or in its early growth. There is also the possibility that the expression of retroviruses is induced as a
consequence of pregnancy. Induction of retroviruses was
reported in mice with an allogenic graft. A similar situation might occur in pregnant women bearing a histoincompatible foetal graft. The first description of C-type virus
particles in human placentas (29) has been confirmed by several groups, and reverse transcriptase activity has been
detected in more than 80 % of full term human placentas
(30). These results, however, do not provide unequivocal
evidence that retroviruses are preferentially expressed
in placental tissues since other organs have not been subjected to comparable studies.

d. Immune responses to retrovirus antigens in pregnant women.

Indications that retrovirus antigens are transiently
expressed during pregnancy came from the findings that
lymphocytes from pregnant women recognized simian retrovirus antigens and reacted in lymphoblastogenesis assays
in vitro, while these reactions were negative in most women with similar numbers of pregnancies, but studied at
least 10 months after the last delivery (31). The incidence of positive responses was greater in multiparous
than in primiparous pregnant women. In other work, 6
women followed during pregnancy developed cell-mediated
reactivity against retrovirus infected cells, as assayed
by lymphocytotoxicity. Responsiveness peaked during the
second and third trimesters of pregnancy and corresponded
with elevated levels of antibodies to the same retrovirus
(32).

e. Specificity of retrovirus markers related to human reproduction.

At this point, it is necessary to comment on the retrovirus antigens detected by the various investigators. Reverse transcriptase from human sperm was specifically neutralized by an antiserum to the BeV enzyme. In contrast,
reverse transcriptase from full term placentas was not
susceptible to BeV antibodies, and the in vitro enzyme
activity required Mg^{++} instead of Mn^{++}, an indication that
this enzyme was associated with a D-type virus. Apparently,
susceptibility of the placental enzyme to anti MPMV antibodies was not tested. In pregnant women, transient lymphocytotoxicity was directed towards cells infected with BeV,
and the antibodies precipitated BeV antigens. These tests
were not performed with MPMV antigens. Lymphoblastogenesis
was induced in vitro by either BeV or MPMV infected cells.
Although these two viruses are very different from each
other, they share one antigenic determinant . However, in
the lymphoblastogenesis assays, women who reacted to BeV
antigens did not always respond to MPMV. No reactions
against SiSV antigens were detected either by lymphoblasto-

genesis .or by lymphocytotoxicity assays.

f. Immune reactions to retrovirus antigens in cord blood.

Our unpublished studies show that :

1. In pregnancies where maternal lymphocytes were significantly stimulated in vitro by either BeV or MPMV antigens, lymphocytes from the cord blood reacted to the same viral antigen as those of the mother. There were some additional cases where positive results were obtained with the infant's blood, but the mothers lymphocytes did not respond. The latter cases occurred mostly in first pregnancies. It was also noted that the indexes of lymphocyte stimulation were higher in cord rather than in maternal blood in first pregnancies. These results indicated that the retrovirus antigens which sensitized the lymphocytes were located at a site available for recognition by both foetus and mother, and that the degree of sensitization of the mothers increased with the number of pregnancies without a corresponding increase in successive foetuses.

2. Neutralizing antibodies to BeV assayed by inhibition of syncytium formation on KC cells were present in 19 of 35 (54 %) of the maternal sera but only in 8 of 35 (23%) of the cord blood sera. In addition, cord sera showed lower neutralizing activity than the maternal sera as determined by the amount of syncytium forming units neutralized by sera at final dilution 1 : 10.

3. Pairs of maternal and cord sera contained complement dependent cytotoxic antibodies specific for ^{51}Cr labelled dog thymus cells producing BeV, or HeLa and rhesus foreskin cells (940C3) producing MPMV, or both. These antibodies were also found more frequently in maternal than cord serum, and the activity of maternal sera was also greater, as detected by the percentage of ^{51}Cr released by sera at final dilutions of 1 : 20 and 1 : 200.

Two possible explanations suggested themselves. The maternal antibodies could have been of the IgM class which cannot cross the placental barrier. Alternatively, maternal IgG antibodies could have been partly adsorbed while filtering through the barrier which apparently can express retrovirus antigens. The placenta has been shown to function as an immunoadsorbent for antibodies directed against paternal HLA antigens. Experiments indicated that both hypotheses were correct. Thirty five maternal sera with cytotoxic antibodies to BeV infected cells were adsorbed 2 times with a suspension of Staphylococcus Cowan A. Thirteen samples retained activity after this adsorption, indicating the presence of IgM or IgG_3 molecules, which do not bind to protein A of the staphylococcus. To distinguish between IgM and IgG_3, the sera were ultracentrifuged in sucrose gradients. Most of cytotoxicity activity against BeV infected cells was found in the 19S region in 13 of the 35 maternal sera. This would explain the lower titers in the related cord sera in these cases. The other 22 mothers, whose antibodies were of the G_1, G_2 or G_4 immunoglobulin subclasses which can filter through the placenta, passed only a limited amount of anti-BeV antibody to the foetus. This would support the idea that retrovirus antigens expressed by the placenta acted as immunoadsorbents.

We then eluted immunoglobulins from the washed tissue of 17 placentas by treatment at pH2. After concentration, the immunoglobulins preparations were assayed for toxic or blocking activity on dog cells, infected or not with BeV. One preparation was toxic for dog cells in the presence of complement, one was specifically toxic for BeV infected dog cells. Among the 15 non cytotoxic immunoglobulin preparations, 6 (40%) blocked the cytotoxic activity of anti-BeV serum on BeV producing dog cells. It has been shown that antibodies eluted from the placentas show specificities for the trophoblasts. The localized expression of retrovirus antigens in placentas, during pregnancy, may at least partly determine the specificity of the antitrophoblast antibodies. Blocking antibodies to retrovirus antigens may help to protect the foetal tissues against maternal immune response.

g. Retroviruses in pre-eclampsia.

The tests described above were then applied to 28 women with symptoms of pre-eclampsia.

As compared to healthy pregnant women, this group of patients had a higher frequency both of lymphoblastogenic responses to BeV or MPMV antigens (33), and IgM maternal cytotoxic antibodies to retrovirus infected cells (our unpublished results). To determine whether these features are due to an increased expression of retrovirus antigens, or to hyperimmune responses, attempts are under way to detect, localize and quantitate retrovirus antigens which may be present on syncytiotrophoblasts or on the underlying basement membrane. It also appeared that passage of retrovirus antibodies from the mother to the foetus was greater in pre-eclampsia. This might have been due to injury of the choriodecidual junction.

h. Retroviruses, chronic nephritis and kidney graft.

Because retrovirus expression had been described in lymphocytes of mice bearing skin allografts, we attempted to isolate a virus from patients with kidney grafts by cocultivating their leukocytes with SIRC cells (34). In one case, a virus population with dual properties was obtained, those of MPMV, plus those of another virus which could not be typed serologically but which formed plaques on XC cells. It could not be determined if the factors which made the isolation possible were related to the nephritis for which the patient was treated, to activation by the kidney graft or to immunosuppressive treatments. In an attempt to distinguish among these possibilities, sera were obtained from 44 patients prior to renal transplantation, and from 27 of these patients after grafting (35). Neutralization of MPMV syncytium forming units was found in the pre-operative sera of 12 of 24 (50 %) patients with chronic glomerulonephritis but only in 2 of 20 (10 %) of the patients with other renal diseases. The difference between the two groups of patients did not seem to be due to

160

differences of treatment. In particular, there was no correlation between the number of transfusions received and
the incidence of MPMV antibodies. However, seroconversions
were observed after renal transplantation and immunosuppressive treatments. Of 19 patients without MPMV antibodies
before graft, 10 (53 %) acquired these antibodies within
2-10 months after the graft. BeV antibodies were found in
6 % of the patients with chronic glomerulonephritis and in
9 % of those with other diseases. Seroconversions were in
frequent after treatment.

Indications that chronic glomerulonephritis in humans is
associated with the expression of retrovirus antigens were
also obtained by two other groups.

Postmortem studies of three cases of systemic lupus erythematosus showed that immunoglobulins eluted from the
kidneys reacted specifically with the p30 of RD114 virus,
which is related to BeV (36). These results fitted well
with the report of antigens related to HeL-12 virus in the
kidneys and other tissues of a patient who died from SLE
(37). As already stated, HeL-12 virus populations appeared to be made of BeV and SiSV components, and the p30 of
BeV shares some antigenic determinants with MPMV. Because
viral gp proteins were not studied by either of the two
groups just cited the possibility has not been ruled out
that chronic glomerulonephritis is associated with viruses
containing a p30 related to MPMV and BeV, and enveloped by
glycoproteins with MPMV specificities. This could explain
why neutralizing antibodies to MPMV, rather than to BeV,
were found in association with some cases of chronic glomerulonephritis.

I thank doctor RC Hard for helping in the preparation
of this manuscript.

REFERENCES

1. Benveniste, R E , M M Lieber, D M Livingston, C J
   Sherr; Infectious C-type virus isolated from a baboon
   placenta. Nature 248 : 17-20, 1974.

2. Kalter S S , R J Helmke, M Panigel, R L Heberling, P
   J Felsburg, L R Axebrod : Observations of apparent C-
   type particles in baboon (Papio cynocephalus) placentas.
   Science 179 : 1332-1333, 1973.

3. Lemons R S , W G Nash, S J O'Brien, R E Benveniste,
   C J Sherr : A gene(BeVi) on human chromosome 6 is an
   integration site for Baboon type C DNA provirus in human
   cells. Cell 14 : 995-1005, 1978.

4. Donehower L , F Wong-Stoal, D Gillespie : Divergence of Baboon endogenous type C virogenes in primates : genomic viral RNA in molecular hybridization experiments. J. Virology : 932-941, 1977.

5. Fine D , G Schochetman : Type D primate retroviruses : a review. Cancer Res. 38 : 3123-3139, 1978.

6. Theilen G H , W Gould, M Fowler, D Dungworth : C-type virus in tumor tissue of a woolly monkey (Lago-thrix spp.) with fibrosarcoma. J. Natl. Cancer Inst. 47 : 881-889.

7. Wolfe L G , F Deinhardt, G H Theilen, H Rabin, T Kawakami, L K Bustad : Induction of tumors in marmoset monkeys by simian sarcoma virus, type 1 (Lagothrix) : preliminary report. J. Natl. Cancer Inst. 47 : 1115-1120, 1971.

8. Nooter K : Studies on the role of RNA tumour viruses in human leukaemia. In : Publication of the Radiobiological Institute of the organization for Health Research TNO, Rijswijk, The Netherlands, 1979.

9. Gallagher, R E , R C Gallo : Type C RNA tumor virus isolated from cultured human acute myelogenous leuke-mia cells. Science 187 : 350-353, 1975.

10. Okabe H , R V Gilden, M Hatanaka, J R Stephenson, R E Gallagher, R C Gallo, S.R. Tronick, S.A. Aaronson. Immunological and biological characterization of type C viruses isolated from cultured AML cells. Nature 260 : 264-266, 1976.

11. Wong-Staal F , D Gillespie, R C Gallo : Proviral se-quences of baboon endogenous type C RNA virus in DNA of human leukaemic tissues. Nature 262 : 190-195.

12. Nooter K , A M Aarssen, P Bentvelzen, F G de Groot, F G Van Pelt : Isolation of infectious C-type oncorna-virus from human leukaemic bone marrow cells. Nature 256 : 595-597, 1975.

13. Nicolson M O , R V Gilden, H Charman, N Rice, R Heberling, R M Mc Allister : Search for infective mam-malian type-C virus related genes in the DNA of human sarcomas and leukemias : Int. J. Cancer 21 : 700-706 (1978).

14. Sherr C J , G J Todaro : Primate type C virus p30 an-tigen in cells from humans with acute leukemia. Science 187 : 855-857, 1975.

15. Strand M , J T August : Type C RNA virus gene expres-sion in human tissue. J. Virol. 14 : 1584-1596, 1974.

16. Stephenson J R , S A Aaronson : Search for antigens and antibodies cross-reactive with type C viruses of the woolly monkey and gibbon ape in animal models and in humans. Proc. Nat. Acad. Sci. USA 73 : 1725-1729, 1976.

17. Zurcher C , J Brinkhof, P Bentvelzen, J De Man : C-type virus antigens detected by immunofluorescence in human bone tumour cultures. Nature 254 : 457:459,1975.

18. Smith H S , J L Riggs, E L Springer : Expression of antigenic cross-reactivity to RD114 p30 protein in a human fibrosarcoma cell line. Proc. Nat. Acad. Sci.USA 74 : 744-748, 1977.

19. Jacquemin P C , C Saxinger, R Gallo : Surface antibodies of human myelogenous leukaemia leukocytes reactive with specific type-C viral reverse transcriptase. Nature 276 : 230-236, 1978.

20. Snyder H W , T Pincus, E Fleissner : Specificities of human immunoglobulins reactive with antigens in preparations of several mammalian type-C viruses. Virology 75 : 60-73, 1976.

21. Kurth R , N M Teich, R Weiss, R T D Oliver : Natural human antibodies reactive with primate type-C viral antigens. Proc. Nat. Acad. Sci. USA 74 : 1237-1241,1977.

22. Stephenson J R , S A Aaronson : Search for antigens and antibodies crossreactive with type C viruses of the woolly monkey and gibbon ape in animal models and in humans. Proc. Nat. Acad. Sci. USA 73 : 1725-1729, 1976.

23. Charman H P , R Rahman, M H White, N Kim, R Gilden : Radioimmunoassay for the major structural protein of Mason-Pfizer monkey virus : attempts to detect the presence of antigen or antibody in humans. Int. J. Cancer 19 : 498-504, 1977.

24. Devare S G , L O Arthur, D L Fine, J R Stephenson : Type D oncornaviruses : immunologic cross-reactivity between major structural proteins of New and Old world primate virus isolates. J. Virol. 25 : 797-805, 1978.

25. Witkin S S , P J Higgins, A Bendich : Inhibition of reverse transcriptase and human sperm DNA polymerase by anti-sperm antibodies. Clin. Exp. Immunol. 33 : 244-251, 1978.

26. Liebermann D , L Sachs : Co-regulation of type C RNA virus production and cell differentiation in myeloid leukemic cells. Cell 15 : 823-835, 1978.

27. Moroni C , G Schumann : Mitogen induction of murine C-type viruses IV. Effects of lipoprotein E. Coli, pokeweed mitogen and dextran sulphate. J. Gen. Virol. 38 : 497:503, 1978.

28. Panem S , E V Prochownik, W M Knish, W H Kirsten : Cell generation and type C virus expression in the human embryonic cell strain HeL-12. J. Gen. Virol. 35 : 487:495, 1977.

29. Kalter S S , R J Helmke, R L Heberling, M Panigel, A K Fowler, J E Strickland, A Hellman : C-type particles in normal placentas. J. Natl. Cancer Inst. 51 : 225-233, 1974.

30. Nelson J. J A Leong, J A Levy : Normal human placentas contain RNA-directed DNA polymerase activity like that in viruses. Proc. Natl. Acad. Sci. USA. 75 : 6263-6267, 1978.

31. Thiry L , S Sprecher-Goldberger, M Bossens, F Neuray:
    Cell-mediated immune response to simian oncornavirus
    antigens in pregnant women. J. Natl. Cancer Inst. 60 :
    527-532, 1973.

32. Hirsch M S , A P Kelly, D S Chapin, T C Fuller, P H
    Black, R Kurth : Immunity to antigens associated with
    primate C-type oncoviruses in pregnant women. Science
    199 : 1337-1340, 1978.

33. Thiry L , S Sprecher-Goldberger, M Bossens, F Neuray:
    Immune response to primate oncornaviruses in pre-eclamp-
    sia. Lancet 1 : 1268, 1978.

34. Thiry L , S Sprecher-Goldberger, P Vandenbussche, M
    Bossens, P Vereerstraeten, O Hestermans-Medard : Ma-
    son-Pfizer like virus in kidney grafted patients. Med.
    Microbiol. Immunol. 165 : 255-269, 1979.

35. Thiry L , S Sprecher-Goldberger, M Bossens, J Co-
    gniaux-Le Clerc, P Vereerstraeten : Neutralization of
    Mason-Pfizer virus by sera from patients treated for
    renal diseases. J. Gen. Virol. 41 : 587-597, 1979.

36. Mellors R C , J W Mellors : Type C RNA virus-specific
    antibody in human systemic lupus erythematosus demons-
    trated by enzymoimmunoassay. Proc. Natl. Acad. Sci.USA
    75 : 2463-2467, 1978.

37. Ordonez N G , S Panem, A Aronson, H Dalton, A I
    Katz, B H Spargo, W H Kirsten : Viral immune comple-
    xes in Systemic Lupus Erythematosus. C-type viral com-
    plex deposition at extrarenal sites. Virchows Arch.
    B. Cell Path. 25 : 355-366, 1977.

DISCUSSIONS OF PAPERS BY R.A. WEISS AND L. THIRY

Dr Lachmann asked whether the 'sarc' gene was still recognised. It was, and coded for a protein of 60,000 daltons, was concerned with transforming cells in culture and with sarcomagenesis, acting as a protein kinase. Sites of mutation in the 'sarc' gene had been precisely located. The 'sarc' gene presumably had some normal function but probably gave rise to oncogenesis when presented in an unscheduled form.

Following this, the role of the oncornaviruses (oncoviruses) in nature was scrutinised. Their natural role was, presumably, in immunology and development. In red jungle fowl, 'en' and 'gag' genes were expressed in embryos; but in grey jungle fowl, these were not detected and the animals remained healthy. It seemed, therefore, that other, as yet undiscovered, oncornaviruses might be present in the grey jungle fowls. Todaro had suggested that, in some strains of mice, up to 7% of the coding sequences represent exogenous retroviruses. These viruses form glycoproteins in plasma and are found in semen. There have been reports of a gene which suppresses oncornavirus production but little more was known. Such genes might act at many stages and there were many species of oncornavirus in mice.

Dr Weiss concluded this discussion by stressing that the role of the oncornaviruses in nature was a chicken and egg argument all the time.

VISNA. THE BIOLOGY OF THE AGENT AND THE DISEASE

G. PÉTURSSON, J. R. MARTIN, G. GEORGSSON, N. NATHANSON and   AND
P. A. PÁLSSON

1. INTRODUCTION

In the years 1937-1952 a disease of the central nervous
system of sheep was observed in the southwestern part of
Iceland. This disease was called visna which means
wasting or withering. Visna usually presented as a
paralysis of the hind legs, and the course was rather slow
but usually progressive, ending in death (1,2).

Visna was associated with a progressive pneumonia
called maedi (3), and studies in Iceland proved both
syndromes to be caused by closely related viruses (4,5).
Investigations of these diseases and other sheep diseases,
such ar rida, the Icelandic form of scrapie, led
Sigurdsson to formulate his concepts of a special category
of infectious diseases, the slow infections (6,7). A
particular feature of the pathological picture in natural
visna, rather conspicuous demyelinated areas in the
central nervous system, suggested that visna might provide
insights into the demyelinating disorders of man, such
as multiple sclerosis.

Maedi and visna have been eradicated from Iceland
by campaign of slaughtering and restocking, which is a
rather unique example of eradication of a slow virus
disease (3). Work on experimental visna produced by
intracerebral virus injection continues in several laboratories in the world, since many aspects of this slow
infection continue to pose intriguing questions such as:
how does this non-oncogenic retrovirus persist and how
does it produce lesions of the central nervous system (8)?

This paper will not attempt to deal with all details

of the published work on visna. Previous reviews will be useful as sources of further information (3,9,10,11,12,13, 14,15).

2. STRUCTURE OF VISNA VIRUS

2.1. *Ultrastructure of virions*

Electron microscopy of visna-infected tissue culture cells shows that virions are formed by budding from the cytoplasmic membrane (16,17,18). Crescent-shaped budding structures form a particle with a relatively electron - lucent center. This structure appears to condense and to form a smaller (80-120 nm) particle with a central dense nucleoid (30-40 nm), which is believed to represent the fully formed infectious virion. At times an internal membrane can be discerned between the central nucleoid and the outer membrane (19). By negative contrast the visna virions can be seen to be covered by knobs about 10 nm in length (20,21).

The morphology of visna virions is typical for C-type RNA tumor viruses except that the dense nucleocapsid material seems to be closer to the viral membrane during the budding process. Inside the cytoplasm of infected cells structures resembling multilayered budding crescents may sometimes be seen (22). It is not known if they represent incomplete or abortive forms of virus.

2.2. *Physicochemical properties*

Visna virus is readily inactivated by ethyl ether, chloroform, formaldehyde, ethanol and phenol as well as by trypsin treatment. It is stable at -50°C for months and is not markedly inactivated by several cycles of freezing and rethawing. In medium containing 1% sheep serum 90%

of infectivity is lost after 4 months at 4°C, 9 days at
20°C, 24-30 hours at 37°C and 10-15 minutes at 50°C. The
infectivity is relatively stable at pH values between 5.1
and 10 (23).

One of the earliest indications of similarities to RNA
tumor viruses was the finding of Thormar that visna virus
was relatively resistant to inactivation with ultraviolet
irradiation (23). This resistance was similar to that
exhibited by avian oncornaviruses. Further similarities
to this group of viruses are the isopycnic density of visna
virus, 1.15-1.16 g/ml in sucrose, the sedimentation co-
efficient about 600 S and the isoelectric point, 3.8 (11).

2.3. *Nucleic acid*

The major nucleic acid component of visna virions is a
60-70 S species of single-stranded RNA cosedimenting with
the major RNA component of Rous sarcoma virus and co-
migrating with this molecule in polyacrylamide gel electrophoresis. A minor RNA component (4-7 S) is also present
in purified virus preparations (24,25,26,27).

The behaviour of the 70 S RNA of visna with alterations
of ionic strength speaks for considerable secondary
structure of the molecule. It can be dissociated by heat
into two 36 S subunits (28).

The molecular weight of the 70 S RNA of visna has been
estimated to be about 10^7 daltons from sedimentation data
and from direct measurements of length (9.3 μm) by
electron microscopy of the RNA released from virions and
prepared by the Kleinschmidt procedure (29). Shorter
pieces of linear RNA (3.2 μm) seem to correspond to the
36 S subunits with a molecular weight of $3x10^6$ daltons.
Recent studies on retroviruses have led to the conclusion
that the genome consists of two 30-40 S subunits with
molecular weight of about $3x10^6$ daltons each (30). It
seems likely that the visna genome will be shown to

contain two subunits also.

The base composition of the 70 S RNA of visna has been determined as C,A,G,U : 28,23,23,26, (27). Long parts of the molecule (100-200 nucleotides) consist of poly- adenylic acid (Poly A) and these Poly A sequences have been located at the 3´ end of the molecule just as in Rous sarcoma virus (31,32,33).

Thus, the 60-70 S RNA seems to be formed from 2 sub- units of about 10.000 nucleotides each. Some contradictory evidence has been published as to whether these subunits contain identical nucleotide sequences (polyploid model). Transfection with visna proviral DNA apparently required more than one subunit since the kinetics were two-hit (34). Chemical analysis of nucleotides following partial digestion with T1 RNase indicates a genetic complexity of about 10.000 nucleotides, supporting the polyploid model (33,35). It has recently been shown (33) that 19 large RNase T1-resistant oligonucleotides are arranged in the same linear order within all subunits. There are therefore no large redundant segments in the visna virus genome. The methods used ruled out circular permutations, but relative- ly short terminal redundancies could not be excluded. Thus, it appears that the 36 S subunits are largely identical and that the visna genome is therefore polyploid (probably diploid).

2.4. Virus proteins

The structural proteins of visna virus have been analyzed by gel electrophoresis and by chromatography. The number and distribution of visna polypeptides are quite similar to those of avian tumor viruses except for certain differences in the glycopeptides. A major virus component (40% of the virion mass) of 25.000 daltons (p25, sometimes referred to as p30) seems to be a constituent of the virion core and the glycoprotein (gp135) is thought to be

associated with the surface knobs. These two proteins
form the two major precipitin lines in gel immunodiffusion
tests with sera from naturally and experimentally infected
and hyperimmunized sheep.

The number of polypeptides has been reported as 10-25
(36,37,38,39). Since the combined molecular weight of the
polypeptides requires a larger coding capacity than provided by a genome of 10.000 nucleotides (36 S RNA subunits)
it has been proposed that the virion polypeptides are not
all primary gene products but may result from overlapping
proteolytic cleavage of a precursor protein molecule (11).
Of course the complete absence of host proteins is difficult to ascertain even with purified virus preparations.
Much remains to be done in analyzing and characterizing the
protein components of visna and in determining their
structural and functional significance.

The envelope of visna virus has been reported to contain
neuraminic acid. Visna virus will inhibit hemagglutination
by influenza virions and this hemagglutination inhibition
activity can be abolished by treating visna virions with
neuraminidase. Such treatment did not influence attachment
to sheep cells, infectivity nor cell-fusing activity of
visna virus (40).

2.5. *Reverse transcriptase*

Shortly after the discovery of RNA-directed DNA polymerase
or reverse transcriptase in avian and murine oncornaviruses
a similar enzyme was reported in visna virus (41,42,43).

The polymerase of visna catalyses both the formation
of single stranded DNA from an RNA template and subsequent synthesis of double stranded DNA. Several years
earlier Thormar (44) reported the early phases of visna
virus replication to be sensitive to inhibition of DNA
synthesis by 5-bromodeoxyuridine and that actinomycin
D decreased virus yield. Lin and others have reported

three different polypeptides with polymerase activity
associated with visna virus (45). Their polymerase I was
reported to have a molecular weight of 125.000 daltons and
to be similar to reverse transcriptase purified from
oncornaviruses. Whether or how these three enzymes may be
related is not clear.

3. REPLICATION OF VISNA VIRUS

3.1. Virus-cell interactions in vitro

Visna virus differs from many retroviruses in being cytopathic in cultured cells (46). Infected cultures show
both rounded individual cells and multinucleated syncytia.
This leads to destruction of the cell monolayers and
appearance of progeny virus in high titers in the cell
culture medium (10^7-10^8 TCD_{50} per ml).

High input multiplicity visna virus causes direct
cell-fusion from without as early as 30-60 minutes after
inoculation. This effect does not depend on virus multiplication since virus inactivated by ultraviolet light
will produce it. Cells that do not support virus multiplication such as BHK21-F cells may be susceptible to virus-
induced cell fusion (47).

Adsorption of virus to cells is completed in about 2
hours and virus nucleocapsids enter the cell after fusion
of the virus membrane to the cytoplasmic membrane of the
cell. The latent period before newly formed infectious
virus begins to appear is about 16-24 hours (48). This is
followed by an exponential increase of infectious virus
for 20-30 hours after which virus production levels off.

After the virus has entered the cell the viral RNA is
transcribed into DNA (49). Compounds that inhibit DNA
synthesis, especially 5-bromodeoxyuridine, inhibit virus
synthesis but only at early stages, within 8 hours after
infection (44). By transfection experiments (34), it

has been proven that all the necessary genetic information
for the synthesis of complete visna virions is transcribed
from RNA to DNA in the infected cell. It was shown by
Haase and Varmus that at least some of the DNA provirus is
covalently integrated into host cell DNA (49).

Synthesis of viral DNA is thought to take place only in
the nucleus. Synthesis of viral RNA takes place in the
nucleus with subsequent transport into the cytoplasm (50).
By immunofluorescent staining, viral antigens were first
detected in the perinuclear cytoplasm but as budding of
virus proceeds staining for antigens becomes brighter at
the cell surface (51,52).

No visna-virus specific sequences have been detected by
hydridization techniques in uninfected sheep cells (49,50).
It seems clear that visna is an exogenous virus of sheep
that spreads mainly horizontally from one animal to another
as does bovine leukosis virus of cattle (53). This is
of course consistent with the epizootic behaviour of
visna-maedi disease in Iceland and its successful eradication.

Although visna virus is usually grown in cultivated
sheep choroid plexus cells, sheep cell cultures from other
organs of sheep seem to be quite permissive for virus
replication. Cells from certain animal species,
especially bovine cells can support virus multiplication
(54,55). Cells from most other species tested have either
failed to support virus multiplication or produced only
minimal titers.

3.2. Possible oncogenic potential of visna virus

Of particular interest is the report (56) that mouse cell
lines were transformed by visna virus. No free infectious
virus was found in these transformed cell lines, but visna
could be rescued from them by cocultivation with
permissive sheep cells. The transformed mouse cells

produced tumors when inoculated into X-irradiated mice.
Tumor-specific transplantation antigens were found but no
crossreaction between various visna-induced tumors (57).
Other workers found the visna genome to be associated with
mouse cells for up to 20 cell passages but only in a form
that could be detected by cocultivation with susceptible
sheep cells. They could not find any evidence for cell
transformation in these cells nor in hamster cells that
produced small amounts of virus for up to 100 days in
vitro (58).

MacIntyre observed morphological changes in visna-
infected human astrocytoma cells and in sheep cells
infected with a strain of virus reisolated from the human
cells (59,60). The morphological alterations were
reminicent of transformation observed in vitro with oncogenic viruses, accompanied by continued virus production
but could not be confirmed as malignant by inoculation
into transplantation compatible sheep (61). In this
context it should be mentioned that no virus-induced tumors
have ever been observed in visna-infected sheep in Iceland,
some of which have been followed for a number of years.

4. CLINICAL MANIFESTATIONS OF VISNA

Following intracerebral injection of visna virus, there is
a clinically silent period of variable length from a few
months up to 8 or 9 years (62). The factors determining
the length of this preclinical period are completely
unknown at present. When the clinical signs appear they
consist of progressive paralysis, especially of the hind
quarters. The rate with which weakness increases is
different from one case to another and sometimes the
disease seems to be stationary for a while. Even if the
animals are unable to rise, they remain apparently alert
and can still survive for weeks if helped to feed and
water. The intervals from onset to death (or prostration

with sacrifice) is usually 1-12 months.

The distribution of incubation periods in two experimental series is shown in figure 1. This figure indicates that the proportion of animals which develop signs can vary, perhaps due to the strain of virus injected. Clearly, some sheep never develop signs even though they are persistently infected as judged by repeated virus isolations and serum antibody response.

Although sheep exhibit no outward signs of disease during the preclinical period, examination of the spinal fluid reveals pleocytosis starting 1-2 weeks after intra-cerebral injection of virus (1,63,64,65). Most of the cells are macrophages or lymphocytes and a few plasma cells are seen (65,66). Polymorphonuclear leucocytes are practically absent. The number of cells in the spinal fluid increases and reaches a maximum about 1 month after infection. Thereafter it gradually declines but often remains somewhat elevated for a long time (figure 2) with later peaks often appearing towards the end when clinical signs start. No clearcut changes are observed in the peripheral blood.

5. PATHOLOGY

In the early papers of Sigurdsson and his coworkers the pathological features of natural and experimentally transmitted visna were described as inflammatory changes accompanied by demyelination (1,2,67). Later studies by Georgsson and others have primarily focused on character-ization of early pathological changes and they have in general confirmed and extended the results of earlier workers (12,63,68). Pathological changes are already quite prominent in many brains as early as 2-4 weeks after intracerebral injection of virus. They consist of inflammatory changes concentrated around the ventricular system and the central canal of the spinal cord, and

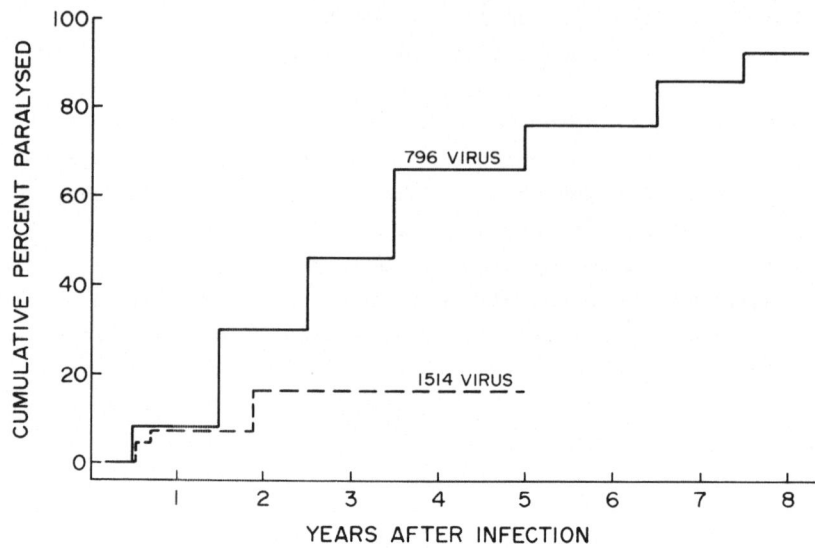

Figure 1.   Cumulative incubation periods in two groups of
            Icelandic sheep injected intracerebrally with
            10^6 TCD_{50} of strains 796 (24 sheep) and 1514
            (20 sheep) of visna virus.   After Pálsson and
            Gudnadóttir (62), and Pétursson et al (63) and
            unpublished.

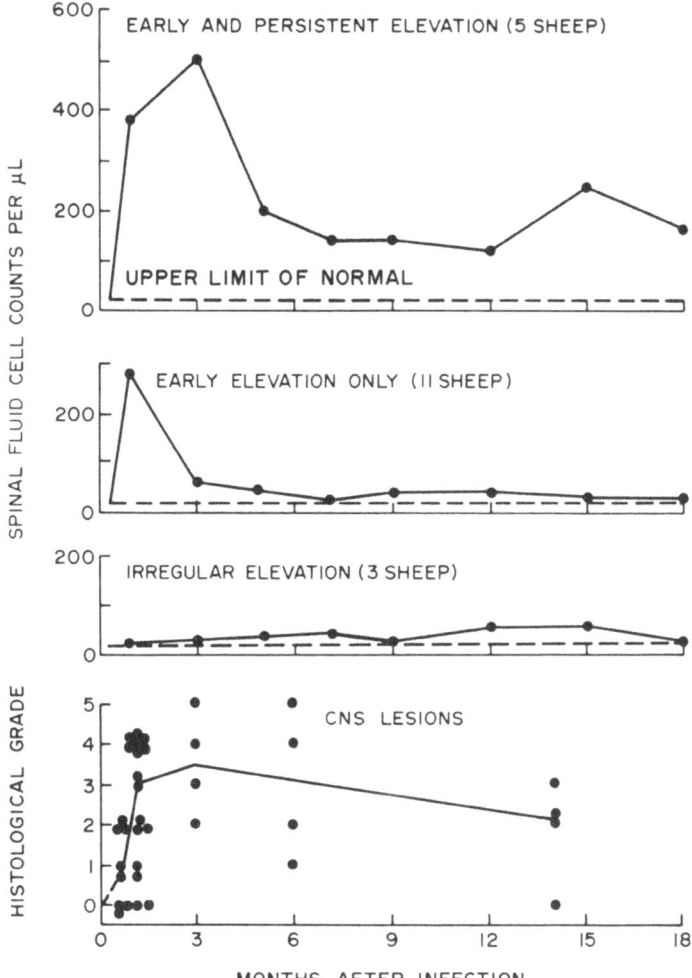

Figure 2. Sequential pathological changes in the CNS of sheep after intracerebral inoculation of 10^6 TCD$_{50}$ of strain 1514 of visna virus. Lower panel: the severity of CNS lesions graded on a scale of 0-6 in 36 sheep. Upper panels: CSF cell counts on a group of 19 sheep which have been sorted into 3 groups to show characteristic patterns. After Pétursson et al (63) and Nathanson et al (65).

extending into adjacent white and grey matter.
Inflammation of the meninges and the choroid plexus is
also a common feature. Perivascular localization of
inflammatory cells is usually observed in visna and glial
nodules are also seen. The cell types involved in the
inflammatory reaction are mainly mononuclear, and polymorphonuclear leucocytes are not seen. Thus the
inflammatory components is similar to that observed in the
viral encephalitides.

In some places the inflammatory changes are quite
severe with destruction of tissue and liquefaction
necrosis and sometimes foci of coagulative necrosis.
Inflammation of the choroid plexus can at times be
particularly intense with massive proliferation of
lymphoid tissue with active germinal centers.

In the early inflammatory lesions evidence of primary
demyelination (destruction of myelin with preservation
of axons) has not been observed. Present investigations
on sheep that have been sacrificed at later stages after
infection indicate, however, that primary demyelination
may be seen in animals with more advanced lesions (66).

In a series of sheep sacrificed two weeks to 13 months
after infection there was little evidence for progression
of lesions with time, except during the first month as
shown in figure 2. There was, however, a definite
correlation between the severity of histopathological
lesions and the frequency of virus recovery (63).

Usually the number of cells in the spinal fluid gives
a rather good indication of the intensity of inflammatory
changes in the central nervous system.

The age of the animals does not seem to have a decisive
influence on the character or severity of lesions according
to comparative studies on the effect of visna virus on
newborn lambs (69) or foetal sheep (70).

6. VIRUS-CELL INTERACTIONS IN VIVO

Visna virus can be isolated from the tissues of infected
sheep beginning 1-2 weeks after infection and at any time
after that (63). Titers of free infectious virus are
usually minimal (table 1) and virus isolation often
requires the use of tissue explants. The success of
isolation depends to some extent on the number of samples
tested and the isolation frequency increases with the
number of methods applied. Often one and occasionally two
blind passages are required for virus isolation (table 2).
  The best source of virus is the central nervous system
especially the choroid plexus but lymphoid organs, spleen,
mesenteric and mediastinal lymph nodes and lungs are
also frequently positive. In the blood the virus is
strictly cell-associated and has never been isolated from
cell-free plasma. The frequency of successful attempts
at virus isolation from buffy coat cells varies
considerably (figure 3). Visna has been repeatedly
isolated from efferent lymph (71); since this represents
a population consisting almost exclusively of lymphocytes,
this suggests that the circulating lymphocyte may carry
the viral genome. In the spinal fluid some free virus can
be found, at least during the first three months after
infection, but after that we have found it to be mainly
cell-associated there as in the blood (63).
  As a result of the low titer of fully formed infectious
virus, typical visna virions have with one exception (66)
not been detected in tissues from infected sheep by
electron microscopy (68). Using immunofluorescence virus
antigens have been very difficult to demonstrate in tissue
sections. On the other hand Haase and coworkers have
demonstrated proviral DNA by in situ hybridization in as
much as 18% choroid plexus cells in vivo (72). They
reported, however, that only 0.025% of these same cells
stained for the p25 antigen by immunofluorescence. By
cloning these choroid plexus cells, 14% of them were

found to produce virus in vitro. Thus the evidence
speaks for a severe restriction of the production of
visna virus in vivo by cells that contain the virus
genome, whereas these same cells become readily permissive
and produce infectious virus when grown in tissue culture.
The restriction in vivo seems to be at the level of
transcription (73).

7. IMMUNE RESPONSE TO VISNA VIRUS INFECTION

7.1. *Interferon*

It has been reported that the replication of visna virus
in permissive sheep choroid plexus cell cultures was
completely unaffected by high concentrations of sheep
interferon induced by polyriboinosinic-polyribocytidylic
acid in fetal lambs or in sheep choroid plexus cultures
(74,75). This interferon inhibited the growth of other
viruses. The resistence of visna to interferon is in
contrast to findings with both avian and murine RNA tumor
viruses, whose growth is blocked by interferon at a late
stage of the replication cycle (76). Cell cultures
persistently infected with visna virus are susceptible to
infection with vesicular stomatitis and vaccinia viruses
indicating that visna virus is a poor inducer of interferon (14,74). It is therefore unlikely that the
restriction in vivo of visna virus is mediated through
interferon.

Table 1

Virus titers of homogenates (before and after low speed
centrifugation) from tissues of visna infected fetal
sheep*. After Georgsson et al (70).

Titer per 0.01 g	Whole Homogenate	Supernate of Homogenate
Negative	0	11
1	24	12
10	6	8
≥100	4	5
≥1000	2	0
Totals	36	36

*Limited to virus-positive specimens.

Table 2

Efficiency of three methods of isolation of visna virus
from tissues of infected sheep*. After Pétursson et al
(63).

Method	Frequency of Isolation Number Percent		Proportion Requiring Blind passage
Homogenization	68/96	71%	53%
Explantation	42/96	44%	84%
Explantation and Cocultivation	47/96	49%	83%

*Limited to virus-positive specimens. Blind passage:
cultures were followed for two weeks and those not
showing cytopathic effect were scraped off glass and
inoculated into a new set of cultures.

180

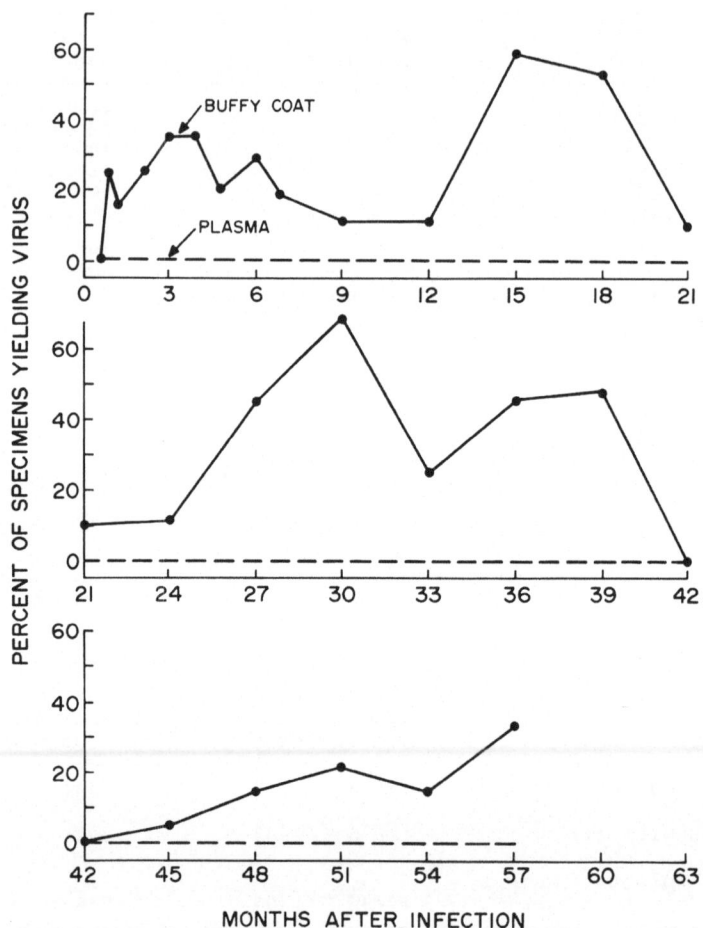

Figure 3. Sequential frequency of isolations of visna
virus from the buffy coat and plasma of 19
sheep inoculated intracerebrally with 10^6
TCD_{50} of strain 1514 of visna virus and tested
over 57 months. Unpublished.

7.2. Virus antibodies in serum

Antibodies directed against virus antigens are induced
following experimental infection of sheep with visna
virus and can be demonstrated in serum by various techniques: complement fixation (77), virus neutralization
(46), immunofluorescence (52), immunodiffusion (78),
passive hemagglutination (79) and as shown recently by
the ELISA technique (80).

Most studies have been done on complement-fixing and
neutralizing antibodies and they have been characterized
to some extent. Complement-fixing antibodies usually
appear about 3-4 weeks after infection but neutralizing
antibodies not until 2-3 months following infection
(figure 4). The titers are fairly high and tend to remain
elevated for years (63,81). Precipitating antibodies
against the glycoprotein appear at 1-6 months in all
infected sheep, while anti-p25 precipitins are slower to
appear (3-24 months) and are only seen in a proportion of
infected sheep. The complement-fixing and neutralizing
antibodies appear both to belong to the IgG1 immunoglobulin class but can be separated on the basis of
differences in electrical charge (8,83). Minimal
antibody activity in the IgM class has been reported (84)
but as yet we have been unable to confirm this.

The complement-fixing antibodies seem to be relatively
nonspecific, not distinguishing between various strains
of visna or even between visna and maedi strains. Undoubtedly, this reflects the fact that the crude antigen
used in complement-fixing tests contains p25 which has
group-specific reactivity. The neutralizing antibodies,
however, are strain specific. Thus some strains of visna
may show little or no crossreaction in neutralization.
The kinetics and optimal conditions for the neutralization
process have been described by Thormar (85).

The target antigens for the antibodies are not as yet
well defined. Neutralizing antibodies are expected to

react with antigens on the virion surface, probably
glycoproteins. In immunodiffusion two major lines can be
detected. One of them represents the p25 core protein
and the other apparently is the major glycoprotein of
the viral envelope.

Antibodies to visna virus can be produced in other
animals than sheep. Thus, after intensive immunization
rabbits will produce complement-fixing, passive
hemagglutinating, immunofluorescent and even neutralizing
antibodies (79). Nonspecific inhibitiors have been found
in low titers in human and in high titers in bovine sera.
They are apparently not immunoglobulins (54,86).

7.3. Virus antibodies in the central nervous system

Neutralizing antibodies to visna virus appear in the
cerebrospinal fluid in some infected sheep. They have
been shown to be produced locally in the central nervous
system (63,64,65). In a few sheep with long-term visna
infection oligoclonal bands in the gammaglobulin region
have been demonstrated by agar gel electrophoresis (65).
Some increase in the protein content of the cerebrospinal
fluid is often observed (1,15), especially of globulins
but not of albumin (64,87).

The presence of plasma cells in the inflammatory
infiltrates and sometimes in the cerebrospinal fluid in
visna infected sheep is consistent with these findings
(66). It is thought that B cell clones generated outside
the central nervous system migrate across the blood-brain
barrier and produce virus-specific antibodies locally in
the central nervous system. Whether this migration is
a chance phenomenon or a directed process is not known
but the presence of visna virus antigen in the central
nervous system probably plays a role. The appearance of
neutralizing antibodies in the cerebrospinal fluid may
well explain the disappearance of free infectious virus

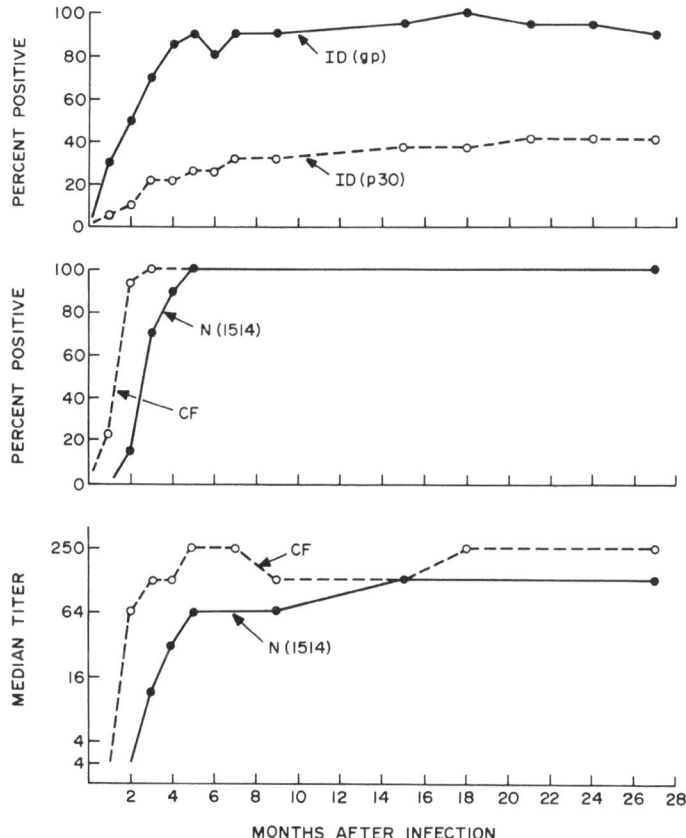

Figure 4.   Sequential comparison of serum antibody
responses of 19 sheep following intracerebral
inoculation of strain 1514 of visna virus.
ID: immunodiffusion test to demonstrate
precipitating antibody against glycoprotein
(gp) and p30 antigens; N: neutralization
test; CF: complement-fixation test. After
Pétursson et al (63) and Nathanson and Gorham
(82).

from this fluid about 3-4 months after infection (63).
Further studies on the antibody response in the central
nervous system are underway and may throw some light
on possible modulating effects of virus antibodies on
the course of the infection and on the pathogenesis of
lesions.

7.4. Antigenic drift

The phenomenon of antigenic drift of visna virus occurring
during the long course of infection in individual animals
was first proposed by Gudnadóttir (9). The appearance of
new antigenic variants that are poorly neutralized by early
antibodies has also been described by Narayan and coworkers
(88,89). Narayan has also reported new antigenic variants
to arise in tissue culture in the presence of neutralizing
antisera and proposed that the appearance of virus variants
under antibody pressure may be a mechanism of persistence
in vivo (90).

7.5. Cell-mediated immune response

Our knowledge of cell mediated immunity to visna virus is
still very scant but some evidence for an early blast
transformation response of circulating lymphocytes to
visna antigens has been found in sheep hyperimmunized with
visna virus (15,71). In intracerebrally infected sheep
the response was transitory, the peak occurred between 1
and 2 weeks after infection both in cells of the cerebrospinal fluid and in the peripheral blood and could no
longer be detected after six weeks (91).

8. PATHOGENESIS

8.1. *Persistence of virus*

From the available data it is clear that visna virus is
rarely if ever completely eliminated from infected sheep
in spite of a fairly good serological response to the
virus and a cell-mediated virus specific response at least
in the early phases of infection. Several mechanisms have
been proposed to explain this failure of the host to clear
the infection.

The most plausible explanation is that the viral genome
persists as a provirus in at least some cells in vivo. If
the genetic information of the virus is not expressed,
there will be no target for viral antibodies or a cell -
mediated immune response. In this way the virus can
escape immune defenses.

The reported insensitivity of visna virus to interferon
and the poor interferon-inducing ability of the virus may
also favor persistence.

The mechanism of antigenic drift may allow new
antigenic variants to escape from strain-specific
neutralizing antibody and thus permit new waves of
infection. A similar phenomenon has been described with
another retrovirus, equine infectious anemia (91). A
systematic study of a group of long-term visna-infected
sheep with careful sequential comparison of virus isolates
has been initiated to determine the importance of antigenic variation in virus persistence.

A defective immune response to visna virus can not be
excluded as a factor contributing to persistence.
Neutralizing antibodies are rather late in appearance and
this may help to establish infection. If our findings
of the absence of an antibody response in the IgM class
can be confirmed this may contribute to persistence. The
inability of the specific immune response to control
multiplication in vivo is evidenced by the observation that

immunosuppression of infected sheep does not lead to
increased virus titers in the tissues at early stages of
infection (93).

8.2. Slowness of clinical evolution

A major factor in the slowness of onset of clinical signs
is the limited permissiveness of sheep cells in vivo. As
already mentioned (63) there is a strong correlation
between the severity of pathological lesions and the
amount of virus that can be recovered from the tissues
(table 5). The restricted expression of the viral genome
may also limit the production of targets for attack by
the immune mechanisms of the host.

So far there is no evidence of the production of
defective interfering virus particles in visna although
such particles have been shown to modify the course of
other virus infections in vivo and in vitro (94).

The late onset of clinical signs might be explained by
a new element in the pathological picture, such as the
development of demyelination. Whatever the mechanism of
tissue destruction it seems obvious that the localization
of lesions must be of crucial importance in producing the
typical paralysis of the hind legs.

8.3. Immunopathogenesis of visna lesions

Several proposals to explain the production of visna
lesions in the central nervous system have been made. In
view of the cytopathic effect (cytolysis and syncytia) of
visna on sheep cells in culture a similar direct effect of
virus in vivo has been proposed (47). Multinucleated
syncytia have, however, not been observed in the central
nervous system in visna (68) as in measles encephalitis.

Indirect effects of visna infection on cells of the

Table 3

Immunosuppression of visna lesions in sheep inoculated intracerebrally with 10^6 TCD_{50} of strain 1514 of visna virus, treated with anti-thymocyte serum and cyclophosphamide, and sacrificed one month after infection. After Nathanson et al (93).

Treatment	Visna Lesion Grade in Individual Sheep	Virus Isolations from CNS
Suppressed	0,0,0,0,0,0,0,2	22/37 59%
Infected Control	1,2,2,3,3,4,4,4	27/39 69%

Table 4

Immunopotentiation of visna, in sheep infected by intracerebral inoculation of 10^6 TCD_{50} of strain 1514 of visna virus, immunized at 3 and 5 weeks with purified virus in CFA, and killed at 8 weeks. Unpublished.

Treatment	Visna Lesion Grade in Individual Sheep	Mean Grade	Percent Positive
Immunized	1,2,2,3,4,4,4,4	3.0*	100%**
Infected Control	0,0,0,0,1,2,3,5	1.4*	50%**

*Not significant   **p <0.01

Table 5

Influence of intracerebral dose on severity of early CNS lesions in sheep inoculated with 1514 strain of visna virus and sacrificed one month later. Unpublished.

Dose (TCD_{50})	Visna Lesion Grade in Individual Sheep	Median Grade	CNS Isolations
$10^{8.5}$	2,3,3,3,3,3,4,4	3.0	22/48 46%
$10^{5.3}$	0,0,1,1,1,1,2,4	1.0	14/48 29%

central nervous system have been suggested. The hypothesis that virus-specific antibodies might react with
visna antigens on the surface of oligodendrocytes and
thus lead to cell lysis and myelin damage (81) has neither
been confirmed nor disproven by direct experimental
evidence. Because it is difficult to find virus particles
by electron microscopy of lesions (68) or to demonstrate
virus antigens by immunofluorescence in tissue sections,
we do not know which cells of the central nervous system
carry viral information or support occasional virus
replication.

We have provided evidence that the lesions of visna,
at least in the early stages of the disease, are immunologically mediated (93). In table 3 it is shown that an
effective immunosuppressive regimen using antithymocyte
serum and cyclophosphamide strikingly suppressed the
development of lesions in visna infected sheep. The
results of an unpublished experiment where we tried to
increase the severity of lesions by immunopotentiation
by injecting visna infected sheep with high doses of
virus in Freund´s adjuvant were consistent with the
notion that visna lesions are immunologically produced
(table 4).

These results raise the question, whether the target
antigens for this immunological attack are viral proteins
or antigens of host tissue origin such as myelin components. To explore this question we compared experimental
allergic encephalitis of sheep to visna by testing for
complement-fixing antibodies to myelin basic protein and
a galactocerebroside antigen in both disorders. In sheep
with experimental allergic encephalitis, antibodies to both
kinds of antigen were produced and a cell-mediated response
to basic protein was demonstrated. No such responses
were found in early stages of visna (95). Therefore it
seems more likely that virus-specified antigens provide
the target for an immunological attack by the host, a view
which is strengthened by the close correlation of severity

of lesions with the amount of virus in the tissues (63).
In recent unpublished experiments we have furthermore shown
that the severity of early lesions in visna can be increased by using a large virus inoculum for infection
(table 5). As mentioned above, autoimmune reactions to
myelin or other host antigens could develop at later
stages of the disease and this possibility is currently
under investigation.

At the moment we view visna as a persistent retrovirus
infection of the central nervous system in which some
cells carry the provirus. Apparently more and more of
these provirus-carrying cells are activated, producing
virus-specified antigens that are attacked by an immune
mechanism, probably a cell-mediated response, with
production of inflammatory lesions. The possibility of
an autoimmune production of myelin damage at later stages
is under study.

9. VISNA-LIKE DISEASE OF GOATS

For some time visna of sheep seemed to be a unique example
of a slow infection caused by a retrovirus and characterized by inflammatory changes and myelin destruction in the
central nervous system. During the last few years reports
have appeared describing a disease in goats which exhibits
many similarities to the clinical picture and pathological
changes of visna. An additional feature of the goat
disease, inflammatory changes of the joints, has been
described both in the U.S.A. and Germany (96,97,98).

From the published observations it seems possible that
the disease syndromes described by German and American
workers are similar although certain differences in
pathology have been stressed (99). The goat disease
seems to run a more acute course than visna, at least in
many instances. A retrovirus has recently been found in
affected goats by several workers (100,101,102). This

agent shows some antigenic relationship to visna virus although it appears to be a distinct virus (103). The opportunity for comparative studies may provide important new insights into slow infections caused by retroviruses.

ACKNOWLEDGEMENT

Supported in part by USPHS grant NS 11451.

REFERENCES

1. Sigurdsson, B, PA Pálsson, H Grímsson: Visna,
    a demyelinating transmissible disease of sheep.
    J Neuropathol Exp Neurol 16: 389-403, 1957.
2. Sigurdsson, B, PA Pálsson: Visna of sheep. A
    slow, demyelinating infection. Br J Exp Pathol
    39: 519-528, 1958.
3. Pálsson, PA: Maedi and visna in sheep. In: Slow
    virus diseases of animals and man, Kimberlin,
    RH, (ed), Amsterdam, North-Holland Publ Co,
    1976, p17-43.
4. Sigurdardóttir, B, H Thormar: Isolation of a viral
    agent from the lungs of sheep affected with maedi.
    J Infect Dis 114: 55-60, 1964.
5. Thormar, H, H Helgadóttir: A comparison of visna
    and maedi viruses II. Serological relationship.
    Res Vet Sci 6: 456-465, 1965.
6. Sigurdsson, B: Observations on three slow infections
    of sheep. Br Vet J 110: 255-270, 1954.
7. Sigurdsson, B: Atypically slow infectious diseases.
    In: Livre Jubilaire du Dr. Ludo van Bogaert,
    Brussels, Acta Medica Belgica, 1962, p738-753.
8. Nathanson, N, H Panitch, G Pétursson: Pathogenesis
    of visna: review and speculation. In: Slow virus
    diseases of animals and man, Kimberlin, RH, (ed),
    Amsterdam, North-Holland Publ Co, 1976, p115-131.
9. Gudnadóttir, M: Visna-maedi in sheep. Prog Med
    Virol 18: 336-349, 1974.
10. Thormar, H, FH Lin, RS Trowbridge: Visna and maedi
    viruses in tissue culture. Prog Med Virol 18:
    323-335, 1974.
11. Haase, AT: The slow infection caused by visna virus.
    Curr Top Microbiol Immunol 72: 101-156, 1975.
12. Georgsson, G, N Nathanson, PA Pálsson, G Pétursson:
    The pathology of visna and maedi in sheep. In:
    Slow virus diseases of animals and man, Kimberlin,
    RH, (ed), Amsterdam, North-Holland Publ Co, 1976
    p61-96.
13. Harter, DH: The detailed structure of visna-maedi
    virus. In: Slow virus diseases of animals and man,
    Kimberlin, RH, (ed), Amsterdam, North-Holland Publ
    Co, 1976, p45-60.
14. Thormar, H: Visna-maedi virus infection in cell
    cultures and in laboratory animals. In: Slow
    virus diseases of animals and man, Kimberlin, RH,
    (ed), Amsterdam, North-Holland Publ Co, 1976,
    p97-114.

192

15. Pétursson, G, N Nathanson, PA Pálsson, J Martin, G Georgsson: Immunopathogenesis of visna. A slow virus disease of the central nervous system. Acta Neurol Scand, (Suppl 67), 57: 205-219, 1978.

16. Thormar, H: An electron microscope study of tissue cultures infected with visna virus. Virology 14: 463-475, 1961.

17. Coward, JE, DH Harter, C Morgan: Electron microscopic observations of visna virus-infected cell cultures. Virology 40: 1030-1038, 1970.

18. Dubois-Dalcq, M, TS Reese, O Narayan: Membrane changes associated with assembly of visna virus. Virology 74: 520-530, 1976.

19. Pautrat, G, P de Micco, J Tamalet: Observation en microscopie électronique du cycle infectieux du virus visna sur cellules de plexus choroides de mouton á des températures inférieures á 37°C. C R Acad Sci [D] (Paris) 283: 211-214, 1976.

20. Thormar, H, JG Cruickshank: The structure of visna virus studied by the negative staining technique. Virology 25: 145-148, 1965.

21. Pautrat, G, J Tamalet, C Chippaux-Hyppolite, M Brahic: Etude de la structure du virus visna en microscopie électronique. C R Acad Sci [D] (Paris) 273: 653-655, 1971.

22. Coward, JE, DH Harter, KC Hsu, C Morgan: Ferritin-conjugated antibody labeling of visna virus. Virology 50: 925-930, 1972.

23. Thormar, H: A comparison of visna and maedi viruses. I. Physical, chemical and biological properties. Res Vet Sci 6: 117-129, 1965.

24. Brahic, M, J Tamalet, C Chippaux-Hyppolite: Virus Visna: Isolement d´une molécule d´acide ribonucléique de haut poids moléculaire. C R Acad Sci [D] (Paris) 272: 2115-2118, 1971.

25. Brahic, M, J Tamalet, P Filippi, L Delbecchi: The high molecular weight RNA of visna virus. Biochimie 55: 885-891, 1973.

26. Harter, DH, J Schlom, S Spiegelman: Characterization of visna virus nucleic acid. Biochim Biophys Acta 240: 435-441, 1971.

27. Lin, FH, H Thormar: Characterization of ribonucleic acid from visna virus. J Virol 7: 582-587, 1971.

28. Haase, AT, AC Garapin, AJ Faras, JM Taylor, JM Bishop: A comparison of the high molecular weight RNA´s of visna virus and Rous sarcoma virus. Virology 57: 259-270, 1974.

29. Friedmann, A, JE Coward, DH Harter, JS Lipset, C Morgan: Electron microscopic studies of visna virus ribonucleic acid. J Gen Virol 25: 93-104, 1974.

30. Bishop, JM: Retroviruses. Annu Rev Biochem 47: 35-88, 1978.

31. Gillespie, D, K Takemoto, M Robert, RC Gallo: Polyadenylic acid in visna virus RNA. Science 179: 1328-1330, 1973.

32. Vigne, R, M Brahic, P Filippi, J Tamalet:
    Complexity and polyadenylic acid content of visna
    virus 60-70 S RNA. J Virol 21: 386-395, 1977.
33. Vigne, R, P Filippi, M Brahic, J Tamalet: Absence
    of circularly permuted and largely redundant
    sequences in the genome of visna virus. J Virol
    28: 543-550, 1978.
34. Haase, A, BL Traynor, PE Ventura, DW Alling:
    Infectivity of visna virus DNA. Virology 70:
    65-79, 1976.
35. Beemon, KL, AJ Faras, AT Haase, PH Duesberg,
    JE Maisel: Genomic complexities of murine
    leukemia and sarcoma, reticuloendotheliosis, and
    visna viruses. J Virol 17: 525-537, 1976.
36. Mountcastle, W, D Harter, P Choppin: The proteins
    of visna virus. Virology 47: 542-545, 1972.
37. Haase, AT, JR Baringer: The structural polypeptides
    of RNA slow viruses. Virology 57: 238-250, 1974.
38. Lin, FH, H Thormar: Substructures and polypeptides
    of visna virus. J Virol 14: 782-790, 1974.
39. Lin, FH: Polyacrylamide gel electrophoresis of
    visna virus polypeptides isolated by agarose gel
    chromatography. J Virol 25: 207-214, 1978.
40. August, MJ, DH Harter, RW Compans: Characterization
    of visna virus envelope neuraminic acid. J Virol
    22: 832-834, 1977.
41. Lin, FH, H Thormar: Ribonucleic-acid-dependent
    deoxyribonucleic acid polymerase in visna virus.
    J Virol 6: 702-704, 1970.
42. Stone, LB, E Scholnick, KK Takemoto, SA Aaronson:
    Visna virus: a slow virus with an RNA dependent
    DNA polymerase. Nature 229: 257-258, 1971.
43. Schlom, J, DH Harter, A Burny, S Spiegelman: DNA
    polymerase activities in virions of visna virus,
    a causative agent of a "slow" neurological disease.
    Proc Natl Acad Sci USA 68: 182-186, 1971.
44. Thormar, H: Effect of 5-bromodeoxyuridine and
    actinomycin D on the growth of visna virus in cell
    cultures. Virology 26: 36-43, 1965.
45. Lin, FH, M Genovese, H Thormar: Multiple activities
    of DNA polymerase from visna virus. Prep Biochem
    3: 525-539, 1973.
46. Sigurdsson, B, H Thormar, PA Pálsson: Cultivation
    of visna virus in tissue culture. Arch Ges
    Virusforsch 10: 368-381, 1960.
47. Harter, DH, PW Choppin: Cell-fusing activity of
    visna virus particles. Virology 31: 279-288, 1967.
48. Thormar, H: The growth cycle of visna virus in
    monolayer cultures of sheep cells. Virology 19:
    273-278, 1963.
49. Haase, AT, HE Varmus: Demonstration of a DNA pro-
    virus in the lytic growth of visna virus.
    Nature [New Biol] 245: 237-239, 1973.
50. Brahic, M, P Filippi, R Vigne, AT Haase: Visna
    virus RNA synthesis. J Virol 24: 74-81, 1977.

51. Harter, DH, KC Hsu, HM Rose: Immunofluorescence and cytochemical studies of visna virus in cell culture. J Virol 1: 1265-1270, 1967.
52. Thormar H: Visna and maedi virus antigen in infected cell cultures studied by the fluorescent antibody technique. Acta Pathol Microbiol Scand 75: 296-302, 1969.
53. Burny, A, F Bex, H Chantrenne, Y Cleuter, D Dekegel, J Ghysdael, R Kettmann, M Leclercq, J Leunen, M Mammerickx, D Portetelle: Bovine leukemia virus involvement in enzootic bovine leukosis. Adv Cancer Res 28: 251-311, 1978.
54. Thormar, H, B Sigurdardóttir: Growth of visna virus in primary tissue cultures from various animal species. Acta Pathol Microbiol Scand 55: 180-186, 1962.
55. Harter, DH, KC Hsu, HM Rose: Multiplication of visna in bovine and porcine cell lines. Proc Soc Exp Biol Med 129: 295-300, 1968.
56. Takemoto KK, LB Stone: Transformation of murine cells by two "slow viruses", visna virus and progressive pneumonia virus. J Virol 7: 770-775, 1971.
57. Law, LW, KK Takemoto: Specific transplantation antigens of murine neoplasms induced by visna and progressive pneumonia viruses. J Natl Cancer Inst 50: 1076-1079, 1973.
58. Brown, HR, H Thormar: Persistence of visna virus in murine and hamster cell cultures without the appearance of cell transformation. Microbios 13: 51-60, 1975.
59. Macintyre, EH, CJ Wintersgill, H Thormar: Morphological transformation of human astrocytes by visna virus with complete virus production. Nature [New Biol] 237: 111–113, 1972.
60. Macintyre, EH, CJ Wintersgill, AE Vatter: Visna virus infection of sheep and human cells in vitro - an ultrastructural study. J Cell Sci 13: 173-191, 1973.
61. Macintyre, EH, G Pétursson, PA Pálsson: Unpublished data.
62. Pálsson, PA, M Gudnadóttir: Unpublished data.
63. Pétursson, G, N Nathanson, G Georgsson, H Panitch, PA Pálsson: Pathogenesis of visna. I Sequential virologic, serologic and pathologic studies. Lab Invest 35: 402-412, 1976.
64. Griffin, DE, O Narayan, JF Bukowski, RJ Adams, SR Cohen: The cerebrospinal fluid in visna, a slow viral disease of sheep. Ann Neurol 4: 212-218, 1978.
65. Nathanson, N, G Pétursson, G Georgsson, PA Pálsson, JR Martin, A Miller: Pathogenesis of visna. IV Spinal fluid studies. J Neuropathol Exp Neurol in press 1979.
66. Georgsson, G, JR Martin, PA Pálsson, N Nathanson, E Benediktsdóttir, G Pétursson: An ultrastructural

study of the cerebrospinal fluid in visna. Submitted for publication.

67. Sigurdsson, B, PA Pálsson, L van Bogaert: Pathology of visna. Transmissible demyelinating disease in sheep in Iceland. Acta Neuropathol 1: 343-362, 1962.

68. Georgsson, G, PA Pálsson, H Panitch, N Nathanson, G Pétursson: The ultrastructure of early visna lesions. Acta Neuropathol (Berl) 37: 127-135, 1977.

69. Pálsson, PA, G Georgsson, G Pétursson, N Nathanson: Experimental visna in Icelandic lambs. Acta Vet Scand 18: 122-128, 1977.

70. Georgsson, G, G Pétursson, A Miller, N Nathanson, PA Pálsson: Experimental visna in foetal Icelandic sheep. J Comp Pathol 88: 597-605, 1978.

71. Martin, JR, G Georgsson, N Nathanson, PA Pálsson, G Pétursson: Unpublished data.

72. Haase, AT, L Stowring, O Narayan, D Griffin, D Price: Slow persistent infection caused by visna virus: role of host restriction. Science 195: 175-177, 1977.

73. Haase, AT, M Brahic, D Carroll, J Scott, L Stowring, B Traynor, P Ventura, O Narayan: Visna: an animal model for studies of virus persistence. In: Persistent viruses, Stevens, JG, GJ Todaro, CF Fox (eds), UCLA symposia on molecular and cellular biology, Vol XI, New York, Academic Press, 1978 p643-654.

74. Trowbridge, RS: Long-term visna virus infection of sheep choroid plexus cells: initiation and preliminary characterization of the carrier cultures. Infect Immun 11: 862-868, 1975.

75. Carroll, D, P Ventura, A Haase, CR Rinaldo, Jr, JC Overall, Jr, LA Glasgow: Resistance of visna virus to interferon. J Infect Dis 138: 614-617, 1978.

76. Pitha, PM, WP Rowe, MN Oxman: Effect of interferon on exogenous, endogenous and chronic murine leukemia virus infection. Virology 70: 324-338, 1976.

77. Gudnadóttir, M, K Kristinsdóttir: Complement-fixing antibodies in sera of sheep affected with visna and maedi. J Immunol 98: 663-667, 1967.

78. Terpstra, C, GF de Boer: Precipitating antibodies against maedi-visna virus in experimentally infected sheep. Arch Ges Virusforsch 43: 53-62, 1973.

79. Karl, SC, H Thormar: Antibodies produced by rabbits immunized with visna virus. Infect Immun 4: 715-719, 1971.

80. Houwers, D: Personal communication.

81. Gudnadóttir, M, PA Pálsson: Host-virus interaction in visna infected sheep. J Immunol 95: 1116-1120, 1965.

82. Nathanson, N, JR Gorham: Unpublished, 1978.

83. Pétursson, G, B Símonarson, B Magnadóttir:
     Characterization of antibodies against visna
     virus. Proc 12th Nordic Veterinary Cong, Reykjavík,
     1974, p252.
84. Mehta, PD, H Thormar: Neutralizing activity in
     isolated serum antibody fractions from visna-
     infected sheep. Infect Immun 10: 678-680, 1974.
85. Thormar, H: Neutralization of visna virus by
     antisera from sheep. J Immunol 90: 185-192, 1963.
86. Thormar, H, H von Magnus: Neutralization of visna
     virus by human sera. Acta Pathol Microbiol Scand
     57: 261-267, 1963.
87. Sigurdsson, B, D Karcher, M van Sande, A Lowenthal:
     Electrophoresis of serum and CSF proteins in
     sheep neurological diseases. In: "Protides of
     the biological fluids", Amsterdam,Elsevier Pub Co,
     1961, p110-111.
88. Narayan, O, DE Griffin, J Chase: Antigenic shift
     of visna virus in persistently infected sheep.
     Science 197: 376-378, 1977.
89. Narayan, O, D Griffin, AM Silverstein: Slow virus
     infection: replication and mechanisms of
     persistence of visna virus in sheep. J Infect Dis
     135: 800-806, 1977.
90. Narayan, O, DE Griffin, JE Clements: Antigenic
     mutation of visna in sheep. J Gen Virol 41:
     343-352, 1978.
91. Griffin, DE, O Narayan, RJ Adams: Early immune
     responses in visna, a slow viral disease of sheep
     J Infect Dis 138: 340-350, 1978.
92. Kono, Y, K Kobayashi, Y Fukunaga: Arc Ges Virus-
     forsch 41: 1-10, 1973.
93. Nathanson, N, H Panitch, PA Pálsson, G Pétursson,
     G Georgsson: Pathogenesis of visna. II Effect
     of immunosuppression upon early central nervous
     system lesions. Lab Invest 35: 444-451, 1976.
94. Huang, AS: Defective interfering viruses. Annu Rev
     Microbiol 27: 101-117, 1973.
95. Panitch, H, G Pétursson, G Georgsson, PA Pálsson,
     N Nathanson: Pathogenesis of visna. III Immune
     responses to central nervous system antigens in
     experimental allergic encephalomyelitis and visna.
     Lab Invest 35: 452-460, 1976.
96. Stavrou, D, N Deutschländer, E Dahme: Granulomatous
     encephalomyelitis in goats. J Comp Pathol 79:
     393-396, 1969.
97. Dahme, E, D Stavrou, N Deutschländer, W Arnold, E
     Kaiser: Klinik und Pathologie einer übertragbaren
     granulomatösen Meningoencephalomyelitis (g MEM)
     bei der Hausziege. Acta Neuropathol (Berl) 23:
     59-76, 1973.
98. Cork, LC, WJ Hadlow, TB Crawford, JR Gorham, RC
     Piper: Infectious leukoencephalomyelitis of
     young goats. J Infect Dis 129: 134-141, 1974.
99. Cork, LC, WJ Hadlow, JR Gorham, RC Piper, TB
     Crawford: Pathology of viral leukoencephalo-

myelitis of goats. <u>Acta Neuropathol</u> (Berl) 29: 281-292, 1974.

100. Weinhold, E: Visna-Virus-ähnliche Partikel in der Kultur von Plexus chorioideus-Zellen einer Ziege mit Visna-Symptomen. <u>Zentralbl Veterinaermed [B]</u> 21: 32-36, 1974.

101. Cork, LC: Personal communication, 1978.

102. Crawford, TB: Personal communication, 1978.

103. Cork, LC, O Narayan: Personal communication, 1978.

EPIZOOTIOLOGY OF MAEDI/VISNA IN SHEEP

G. F. DE BOER AND D. J. HOUWERS

'Maedi' (progressive interstitial pneumonia) and 'visna'
(meningo-leucoencephalitis) are slowly progressive non
febrile contagious diseases of sheep. In the late thirties
both conditions (maedi=dyspnoea and visna=wasting) were
observed for the first time in Iceland and were considered
to be two different diseases. In the same period jaagsiekte
or sheep pulmonary adenomatosis (another lung condition of
sheep) and rida (scrapie) became also apparent in the Icelandic sheep population.
Sigurdsson and his coworkers were studying these diseases
and it became evident that these conditions were characterised by silent but relentlessly progressive lesions which
developed over a long period of time. It was thought that
these slowly progressive diseases could neither be grouped
as acute nor chronic diseases. In acute infections the
disease runs a rather regular course, the causative agent
enters the body where it multiplies and spreads rapidly so
that clinical signs appear after an incubation period of a
few days. The hosts defences are mobilised and unless the
patient dies the infecting agent is eliminated and convalescence begins. Chronic infections on the other hand are not
only much more protracted in their course, they are also
much less regular and unpredictable. In order to character-
ise the group of four sheep diseases, Sigurdsson coined the
term 'peculiarly slow progressive infectious diseases'
(annarlega haeggengir smitsjúkdómar), often spoken of as
'slow virus diseases' (1).
In 1957 a virus was isolated in tissue culture from the
brains of five experimentally transmitted cases of visna.
It was subsequently demonstrated that this isolate was not
only neutralised by sera from sheep with a disorder of the
central nervous system but also by serum samples from sheep
affected with maedi (2). When new maedi outbreaks occurred

D. A. J. Tyrrell (ed.), Aspects of Slow and Persistent Virus Infections, 198-220. All Rights Reserved.

in Iceland, the techniques which had been successful for the
isolation of visna virus were applied to lungs of maedi
sheep and maedi virus became available for experimental
studies (3).

In 1964, in the Netherlands virus was recovered from lungs
of sheep with 'zwoegerziekte', an interstitial pneumonia
resembling maedi (4, 5) and serologically similar viruses
were isolated from brain material of sheep with neurological
disorders and histopathological lesions of the central
nervous system resembling visna (6). Various properties of
these isolates were found to be in accordance with data
reported for maedi and visna viruses (4, 7, 8). Experimental
infections with virus recovered from the lungs of sheep
suffering from zwoegerziekte caused progressive interstitial
pneumonia (maedi) and meningo-leucoencephalitis (visna).
Hence the name maedi/visna virus was proposed for the agent
which causes both disease entities (9). One name for the
causative agent of the disease, which has been described
under various local names, seems well justified since
molecular hybridisation studies (10) have shown homologous
nucleotide sequences in strains of maedi and visna virus.
The name maedi/visna virus gives credit to the Icelandic
investigators who were the first to recognise the aetiology
of both diseases.

CLINICAL SYMPTOMS OF MAEDI AND VISNA

The first symptoms of maedi are usually observed in sheep
older than 3 to 4 years. Illthrift is one of the first signs
that may be observed by an experienced sheep farmer. The
affected animal shows a greyish discoloration of the fleece
and the abdomen is thin. The suspected animals lag behind
and show signs of respiratory distress when the flock is
moved to another pasture. After some weeks, sometimes months,
the increased rate of respiration becomes also manifest
without a preceding physical effort. Despite a good appetite affected animals lose weight and finally become
cachectic. In the terminal phase of disease the animals lie

down most of the time. The entire clinical stage may last
for some months and sometimes for more than a year. The
fatal course of the disease is appreciably accelerated when
the sheep is kept under conditions of stress e.g. during
fostering of lambs. Following intrapulmonary inoculation of
16 sheep of the Texel breed with a zwoegerziekte virus
strain of maedi/visna virus, symptoms were first observed
after 26 to 31 months in four sheep, whereas the longest
interval amounted to 58 months (4). Contact transmission
studies have shown that the duration and degree of exposure
greatly influenced the time of onset of disease, which again
varied from 2 to 5 years (11).

Visna is only observed in sheep over two years of age.
The first sign noted is that the sheep lag behind, inco-
ordination becomes apparent when the flock is moved from one
pasture to another. Meningitis is an early event of visna,
as judged by pleocytosis of cerebrospinal fluid which is
present shortly after experimental infection. Despite a
persisting appetite the animal looses weight. Gradually the
paresis of the limbs progresses and walking becomes difficult. Mostly the hind limbs are affected. Tremors of the
head and facial muscles and blindness are occasionally seen.
The paresis slowly progresses to paralysis, prostration and
death (12).

PATHOLOGY AND PATHOGENESIS

At necropsy overt changes of maedi are confined to the
thoracic cavity, lungs and associated lymph nodes. The lungs
are enlarged, weighing in advanced cases two to three times
as much as normal lungs. The shape is not much altered but
the affected tissue is of firm consistency. The normal
pinkish-red colour of a healthy sheep lung is replaced by a
characteristic greyish colour. The most advanced lesions are
usually found in the diaphragmatic lobes and occur less
frequently in the cardiac and apical lobes. In about 10% of
the cases histological examination is desirable for the
diagnosis (5, 13, 14). Often secondary bacterial infections

complicate the lesions. The microscopic lesions
develop progressively and consist of a thickening of the
interalveolar septa caused by proliferation of alveolar
septal cells and infiltration with mononuclear cells, mainly
lymphocytes, monocytes and macrophages. Another early feature is hyperplasia of peribronchial, perivascular and lymphoid tissue. As the disease advances the cell infiltrates
are replaced by fibroblasts and argentophilic fibres which
turn into collagen fibres. Fibrosis with strands of collagen
fibres may develop in areas with very pronounced thickening
of the interalveolar septa. The process is accompanied by
proliferation of alveolar epithelium and smooth muscular
tissue around the terminal bronchioles. Vascular alterations
which involve the smaller and medium sized arteries are also
common in advanced stages of maedi. Very little is known
about the pathogenesis of maedi. Infectious virus is mainly
recovered from tissue and organs containing lymphoid cells
and in the viremic stage infectivity is associated with
lymphocytes (4).
Macroscopic lesions of visna are only seen after several
weeks of overt clinical symptoms. These consist of emaciation and muscular atrophy. Histopathological lesions of visna
are confined to the central nervous system, where meningeal
and subependymal infiltrations consisting of lymphocytes,
monocytes and some plasma cells are found. Often the infiltrations are small, but in severe cases large areas with
intensive inflammation sometimes accompanied by necrosis are
observed. Around these lesions extensive perivascular cuffs
of lymphocytes, monocytes and a few plasma cells are found
(7, 13, 14). The pathogenesis of visna seems to be immunologically mediated. By applying immunosuppression to sheep
when experimentally infected, Nathanson *et al.* (15) could
demonstrate suppression of the central nervous system
inflammatory response, without apparent effect on virus
replication.
Maedi/visna virus infections of sheep are characterised by
persistence of virus and antibody (16, 17, 18). Only some of
the animals that pass through a stage of viremia and develop

antibodies come down with the disease. Persistent subclinical infections lasting up to 5½ years have been observed
following intrapulmonary inoculation (19). At autopsy, no
macroscopic or microscopic lesions were detected in some
experimentally infected animals, despite the isolation of
virus from the blood at irregular intervals during the
experiment. The conclusion can therefore be drawn that
maedi/visna virus infections do not invariably lead to
clinical disease and histopathological lesions. The concept
of an inevitably fatal course appears to be valid only when
clinical signs have been observed. The long incubation
period and the slow development of clinical disease (slow
virus disease) is thought to be due to a restriction of
virus replication, which should only occur in vivo (20).
The observed ability of the virus to undergo antigenic
modulation (antigenic shift) under the pressure of neutralising antibody was suggested as a second mechanism for the
persistence of maedi/visna virus (21).

SEROLOGY

The persistent infection of maedi/visna virus causes an
immune response of the host which is usually followed by
formation of specific antibodies (16, 17). In an epizootiological study, however, a number of sheep developed histological lung lesions in the absence of detectable amounts
of neutralising, complement fixing or precipitating antibody which was tested at intervals for several years
(11).
For epizootiological studies two techniques for detection
of antibody, the agar gel precipitation test (AGPT) and the
complement fixation test (CFT) proved to be most useful.
When field samples were examined with both tests, more posit-
ives were detected than with each of the tests separately
(16, 22, Table 1). AGPT antigen is prepared by freeze-thawing
sonication and ether treatment of infected sheep choroid
plexus cell cultures. Antigen used in the CFT is prepared of

cell culture supernatant. The test is performed in a microtiter system. AGPT antigen preparations contain a glycoprotein (gp) and a protein fraction of MW 23,000 (p23). In general, serum samples from naturally infected animals show a
gp precipitation line and some samples show a p23 precipitation line as well. In our hands some of the samples, which
showed no gp line and were scored negative, later showed
clear p23 lines in block titrations. So, the outcome of the
test is greatly influenced by the actual concentrations of
the fractions in the antigen preparation and the concentrations of the respective antibodies in the serum samples.
Therefore, the sensitivity of the AGPT can be improved by
testing serum dilutions against dilutions of antigen. The
incongruity between results with AGPT and CFT of different
laboratories and of different techniques in the same laboratory (23) are probably also due to differences in composition of antigen preparations. Variations in gels and testing systems have probably less impact.

Recently, in our laboratory an indirect enzyme-linked
immunosorbent assay (ELISA) has been developed for detection of antibodies to maedi/visna virus. The wells of poly-
styrene microtiter plates are coated with a maedi/visna
virus preparation, which was concentrated and purified by
differential centrifugation. After coating the plates are
rinsed. The test serum samples are diluted in the wells and
the plates are incubated. After rinsing, rabbit anti-sheep
IgG coupled to horseradish peroxidase is added. Subsequently
the plates are incubated, rinsed and the enzyme substrate
5-aminosalicylic acid is added. The tests are read by eye or
OD measurement after one hour.

To compare the sensitivity of the three serological techniques, 494 sheep serum samples from the field were tested
(Table 1). All samples which were positive in AGFT or CFT
were also scored in the ELISA. In addition 57 serum samples
were found positive in the ELISA only.

Table 1. Comparison of Agar Gel Precipitation and Complement
         Fixation tests with an Enzyme-Linked Immunosorbent
         Assay

	AGPT	CFT	AGPT+CFT	ELISA
No positive samples/	88/494	73/494	103/494	160/494
No sheep sera tested				
Percentages positive	17.8	14.7	20.8	30.2

In control tests, serum samples from maedi-free flocks, from
sheep suffering from adenomatosis, from gnotobiotic lambs
and from lambs hyperimmunised against different viruses were
tested and scored negative. The ELISA for maedi/visna virus
antibody is sensitive, specific, can be evaluated objectiv-
ely, needs relatively little antigen and is suitable for
screening of large numbers of sera.

DIFFERENTIAL DIAGNOSIS

Lung adenomatosis (jaagsiekte) has been confused with maedi
for a long time. Clinically the two diseases are difficult
to differentiate. In maedi-affected flocks an occasional
weak and dry cough may be heard. Since the interstitial
pneumonia is not accompanied by excretion of fluid or mucus
in bronchi and trachea, coughing in our opinion is mostly
due to lungworm infestations. Jaagsiekte is associated with
an increased bronchial secretion, which may result in nasal
discharge. Farmers in Iceland called this disease 'vota-
maedi' (=wet maedi). The reported incubation period of lung
adenomatosis is shorter than that for progressive interstitial pneumonia (24). At autopsy adenomatous lungs are
generally not evenly enlarged. In early stages the solitary
bacon like tumours are embedded in tissue of normal consistency.
Histopathologically, the two conditions can be differentiated. Maedi starts off with a proliferative reaction of inter-

stitial tissue and lymphoid tissue, whereas the early
alterations in jaagsiekte occur in epithelium of bronchioles
and alveoli. The solitary adenomatous nodules surrounded by
normal parenchyma are a dominant feature in the early stage.
The presence of layers of cuboidal or low-cylindric cells
lining alveoli or smaller bronchi and the projecting papil-
lary ingrowths are specific for advanced cases of adenomatosis. The alveolar spaces and small bronchioles are often
filled with exudate containing large numbers of leucocytes.
Metastasis in mediastinal lymph nodes and even in other
parts of the body has been reported for jaagsiekte (25, 26),
but have never been observed in maedi.

HOST RANGE

Infections with maedi/visna virus seem to be restricted to
sheep and goats. Low titres of neutralising activity against
maedi/visna virus were observed in serum samples of bovine
and human origin (27), but these do not reside in the gammaglobulin fraction of the blood and probably represent non-
specific inhibitors. Numerous attempts to infect small
laboratory animals with maedi/visna virus have failed (28).
Although virus multiplication was shown in tissue cultures
of bovine and recently also in astrocyte cultures of human
origin (29), there is no indication that maedi/visna virus
possesses infectivity for these species in vivo. Sheep
farming always involved   close human contact. Affected
sheep were often slaughtered on the farm without any pre-
cautions and meat from diseased sheep was always, and still
is, considered suitable for human consumption and marketed
accordingly. Fortunately, despite the extensive human exposure, there is no evidence that maedi/visna infections spread
to man.

GEOGRAPHIC DISTRIBUTION AND INCIDENCE

Maedi or maedi-like pneumonias, although not all verified
by virus isolation, have been reported in sheep in several

European countries, in various parts of the USA, East- and
South-Africa and India. The lung condition has been described under various names such as zwoegerziekte, progressive
pneumonia, Montana sheep disease, disease of Graaff-Reinet
and 'la bouhite' (4, 14).

In 1939, Gíslason observed several diseased Icelandic sheep
showing an interstitial pneumonia. Epidemiological observations indicated that all maedi outbreaks could be traced
back to a few sheep of the Karakul breed which were imported
from Germany in 1933. The disease spread gradually through
large parts of the country and by 1945, when the epizootic
was at its peak, about 60% of the sheep farming districts
were affected. At this time the number of winterfed sheep
had declined from 700.000 to 450.000. In individual flocks
the annual losses could reach 20-30% (14). The disease in
combination with visna, jaagsiekte and scrapie (rida) became
an economic disaster for the Icelandic sheep industry. It
was therefore decided to slaughter all sheep on every farm
within affected districts of the country and to repopulate
these farms later with sheep from unaffected areas. This
heroic eradication programme took almost ten years to
accomplish (1944-1954), but with success. In a few districts
in the western and north-western part of the country, 4 to 7
years after the original flocks were removed, maedi was
observed again and the control programme had to be continued.
The last recurrence of maedi in Iceland was in 1965. The
disease had disappeared before laboratory techniques for
epidemiological studies became available (7).

Other countries were not so fortunate, but only a few
studies were performed on the incidence of maedi/visna virus
infections. In 1971 a serological survey was performed with
a random sample of sheep sera from various areas in the
Netherlands. About 3000 serum samples were collected from
larger flocks and per flock about 10% of the adult sheep
were bled. Serum samples were tested for precipitating and
complement fixing antibodies with AGPT and CFT. Antibody was
detected in about 28% of the sheep sera tested. The majority
of the seropositive sheep were detected with the AGPT and an
additional 2% by the use of CFT. In most flocks at least one

animal with antibodies was present. The data suggested a widespread maedi/visna virus infection in the Dutch sheep population (22).

In the last ten years an increase of maedi pneumonias has been observed in sheep of Denmark, Norway, Sweden and Germany (30, 31). Especially in Denmark and Norway maedi/visna virus infections seem to be increasing, but exact information on the incidence is still lacking. Many infections in these four countries were connected with import or trade of sheep of the Texel breed or their crosses with native breeds. However, in serological studies little attention was given to indigenous breeds. In Germany for example, antibodies to maedi/visna virus were detected in 50% of serum samples from a flock of Merino sheep in which clinical symptoms of maedi had never been observed (32).

Maedi/visna (ovine progressive pneumonia) was found to be prevalent in the major sheep-producing areas of the USA. Recently Cutlip *et al.* (33) collected about 1400 serum samples from cull sheep of 5 to 10 years of age at slaughter plants. Precipitating antibodies to maedi/visna virus were present in more than 40% of samples from midwestern and northwestern states. Virus was recovered from the lungs of 20 to 46% of such sheep from the same areas. In addition, in Idaho, relatively high incidences of precipitating antibody were observed in another serological survey of range sheep (34). The presence of precipitating antibody was largely determined by age, the percentage of positive sheep increasing from 16 in yearlings to 83 in ewes older than 7 years.

Originally, in the late thirties only sporadic cases of visna occurred in Iceland, but annual losses of some farms could reach 10% and in a few flocks the visna mortality exceeded that caused by maedi. After the end of the eradication campaign visna had disappeared from the field. Later, the disease was observed in a few other countries and usually in association with maedi (24). It is noteworthy that in Germany visna was seen by prevalence in sheep of the Merino breed (35).

Maedi/visna in goats seems to be a rare event. In both
Germany and the Netherlands only one flock was observed to
be affected. The German flock suffered from disorders of
the central nervous system (36, 37) and the Dutch goats
from interstitial pneumonia. From both outbreaks viruses
were recovered in tissue culture which were similar to
maedi/visna virus. Since the goats in the Netherlands were
kept in close contact with a zwoegerziekte-affected sheep
flock, it is likely that they had been infected by horizontal exposure. In India, however, maedi/visna infections of
goats seem to be more common. Both interstitial pneumonia
(maedi) and leucoencephalitis resembling visna have been
reported and virus was isolated from lungs (38).

SUSCEPTIBILITY OF DIFFERENT BREEDS OF SHEEP

So far no clearcut data have been obtained for differences
in susceptibility to maedi/visna virus infection between
various breeds of sheep. The heavy losses in Iceland caused
by maedi and visna could be suggestive for a genetically
determined higher susceptibility of the Icelandic sheep, but
Icelandic sheep are maintained under more severe stress than
in other countries. In addition, the relatively high mortality of both conditions may be due to the introduction of
virus in an immunologically virgin population. In our experimental studies with a zwoegerziekte strain of maedi/visna
virus (9), lambs of Icelandic and Texel breeds were employed.
Clinical symptoms of maedi and visna became manifest earlier
in the Icelandic sheep, but these were infected with the
small lungworm (Muellerius capillaris) and the sheep of the
Texel breed were specific pathogen free. Lungworms may have
had an enhancing effect on the course of the disease.
In a few countries certain breeds are predominantly infected. In the Netherlands this is no surprise, since 99% of
sheep in this country are of the Texel breed. In Hungary,
however, the disease has only been described in Merino sheep
(39). In the Idaho survey significant differences were noted
between the percentages of serologically positive sera from

sheep of various breeds within three large flocks (34).
Rambouillets demonstrated a significantly lower incidence
of precipitating antibodies than five other breeds while
such antibodies were more frequently present in Finn crosses
than in the other five.

Very little is known about differences of resistance or
susceptibility of strains or blood lines within the same
breed. During the epizootic in Iceland the impression was
obtained that certain strains within the Icelandic breed
were more resistant than others and crosses between native
ewes and Border Leicester rams appeared to be particularly
resistant (24). Our epizootiological observations indicated
a positive correlation between serologically positive ewes
and positive offspring within the Texel breed. This results
probably from a horizontal (lactogenic) transmission of
maedi/visna virus, but may also be influenced by genetically
determined differences in susceptibility to infection.

EPIZOOTIOLOGY

In 1933, in Iceland, the Karakul sheep were kept in quaran-
tine for two months and thereafter, showing no symptoms of
disease, were sent to 14 farms in different districts of
the country. In retrospect, at least two of the rams carried
the infection of maedi and gave rise to two epizootics in
different parts of the country. It soon became clear that
maedi was a contagious disease with an incubation period of
at least 2 to 3 years. The disease was later successfully
transmitted by direct contact between healthy and diseased
sheep, by contaminating drinking water with faeces from
sheep affected with maedi and by injecting maedi lung
material intrapulmonarily and intravenously (24). Sheep
farming practices in Iceland are conductive to the spread
of maedi/visna virus. All sheep are closely housed on the
farms during winter and the size of the flock is kept as
large as the food supply permits. During the summer months
sheep from different farms roam freely in the hills, but
here the communicability of maedi seems to be low, even in

its clinical stage (24). Stress and the degree of direct
contact seem to be prerequisites for the development of
clinical disease. In the Netherlands, severe losses (up to
15% mortality) are observed in a few larger flocks. In
addition, in the individual animal, parasitic infestation
(lung worms) may enhance the development of clinical pneumonia.

To study the transmission of maedi under natural conditions,
we performed a field trial with four flocks of about 40
sheep which were separated from the parent flock at different times after birth (11). A total of 146 pregnant ewes of
the Texel breed was purchased from 11 farms in the provinces
of Zeeland and Noord-Holland (Isle of Texel) in the Netherlands. The age of the ewes ranged from 2 to 6 years. The
farms were selected on the basis of a 10 to 20% loss of
adult sheep due to zwoegerziekte in the preceeding years.
The sheep were brought together shortly before the lambing
season on an evacuated farm. Over a period of two months a
total of 220 lambs were born from 135 ewes. Lambing was
supervised day and night,and the lambs were assigned by lot
into four groups.

The lambs of group No 1 were delivered into sterile towels
and immediately transferred to a climatic chamber on the
institute's farm. Lambs of group No 2 were left with their
mothers for 9 to 11 hours, were allowed to take up colostrum from the dams, and were thereafter transported to a
second climatic chamber. Both groups were reared artific-
ially for a period of 2 to 3 months and thereafter transferred to separate paddocks of about two hectares each.
Group No 3 was weaned at the age of 6 weeks and was then
transferred to a third farm. Lambs of group No 4 were
allowed the normal contact with the ewes of the parent
flock for a period of one year. Two years after initiating
the trial, the total numbers of sheep present in groups Nos
1, 2, 3 and 4 were respectively 50, 40, 38 and 38. These
flocks were kept under observation for respectively 8, 6,
7 and 4 years. Serum samples were collected from the sheep
of the four flocks twice a year. Results of serological

testing with AGPT, CFT and neutralisation tests of one
sample per year are presented in Table 2.

During the 8 years of observation, no specific antibody to
maedi/visna was detected in serum samples from sheep of
flock No 1. No symptoms of maedi were observed and plasma
clot cultures of various organs of these animals yielded no
virus either. In flock No 2 a low percentage of sheep with
antibody was detected with the AGPT and CFT, whilst the
neutralisation tests remained negative during an observation
period of 6 years. Signs of maedi were only observed in one
five year-old animal. It showed respiratory distress and
was the first sheep of this flock from which virus was
recovered. Thereafter another four yielded virus. At autopsy
lungs of two sheep of this flock weighed over 1000 g. Histological lesions indicative of maedi were observed in the
lungs of three sheep. In contrast to flocks No 1 and No 2,
antibodies to maedi/visna virus were frequently present in
serum samples of flocks No 3 and No 4. The highest score
was observed in flock No 4, in which the number of positive
sera was about three times higher than in flock No 3 (Table
2). Most sheep of flocks No 3 and No 4 which died or had to
be euthanized, suffered from maedi. Flock No 4 was slaughtered in the fourth year because the majority showed emaciation
and laboured breathing. Histopathologic lesions of maedi
were observed in 55% of the sheep and virus was recovered
from 10 of 35 sheep tested. In flock No 3, these figures
were somewhat lower and symptoms of maedi were mainly seen
in the fifth and sixth year of the trial. The total of
maedi/visna virus infected sheep detected at post mortem
by virological, serological and histological examination
in flocks No 1 to No 4 amounted to 0/50(0%), 11/40(28%),
28/37(76%) and 30/37(81%) respectively.

The trial provided evidence for horizontal transmission
of maedi/visna virus. The results obtained with flock No 2
demonstrate that lambs may become infected within 10 hours
after birth. A prolonged contact exposure, however, results
in a more severe infection, which is reflected by the total
number of serologically, virologically and histopathologic-

Table 2. Percentage of serologically positive sheep in the four flocks which were contact-exposed to the parent flock for different periods.

Flock No	Exposure to parent flock	Serological test	Age of sheep (years)							
			1	2	3	4	5	6	7	8
1	none	AGPT	0	0	0	0	0	0	0	0
		CFT	0	0	0	0	0	0	0	0
		NT	0	0		0	0	0		
2	10 hours	AGPT	5	7	3	6	9	11		
		CFT	0	3	3	3	3	7		
		NT	0	0	0	0	0	0		
3	6 weeks	AGPT	23	28	33	55	47			
		CFT		3	26	18	13			
		NT	25	14	16	0	0			
4	1 year	AGPT	67	67	55	64				
		CFT	62	65	56	26				
		NT	63	60	41					

ally positive and of clinically affected sheep observed in the three flocks.

In an earlier study (4), we reported the recovery of maedi/ visna virus from milk of ewes (of flock No 4), one to five months after lambing. The infection of lambs of flock No 2 therefore was presumably via ingestion of colostrum, although fostering by the dam may have contributed. Lambs of flock No 3 were fostered for a period of 6 weeks. The rate of infection in this flock rose to 76% as compared with 28% in flock No 2. Flock No 4 was weaned from the parent flock at 5 to 6 months of age and again contact-exposed between 7 and 12 months. In winter, they were housed every night in a small barn together with the parent flock. The differences in the degree of contact exposure between the two flocks are well expressed in the time of onset of disease and differences in severity of clinical signs. In addition to ingestion of colostrum and milk, the lambs of flocks No 3 and No 4 may have been exposed to maedi/visna virus via the respiratory route or via ingestion of other materials such as faeces.

Recent observations in the field indicate that practically all serologically positive sheep are born from infected ewes. This is highly suggestive of the importance of lactogenic transmission in the epizootiology of maedi/visna, but one has to keep in mind that genetic predisposition may give a similar picture. The horizontal virus spread may probably be much faster when the infection is introduced for the first time in a sheep flock. We recently observed 90% seropositive sheep in a flock of imported Scottish Halfbreds, which had been in contact with maedi/visna infected sheep for 1½ year.

Sigurdsson *et al.* (40) obtained negative results in their experiments designed to test the role of sheep keds (Melophagus ovinus) in the transmission of disease. However, their experiment had to be terminated after 15 months and so the sheep ked remains a possible candidate for transmitting maedi/visna virus. We found no evidence for the transmission of maedi/visna virus via the small lungworm (Muellerius capillaris, 11).

In our abovementioned field trial, no evidence was obtained
for vertical transmission of maedi/visna virus. In addition, plasma clot cultures of various tissues of a total of
30 fetuses from maedi-affected ewes were invariably negative for virus (11). Therefore, since vertical transmission
of maedi/visna virus does not occur, or at least is of no
importance for the epizootiology of the disease, methods
for controlling the disease appear feasible.

CONTROL OF MAEDI/VISNA

The economic losses caused by maedi/visna virus can be
calculated in terms of earning capacity. No therapeutic
methods are available and in the field sheep are usually
culled from affected flocks before the age of 5 years. By
this procedure mortality rates are kept low, but the earning
capacity is not optimal. Control programmes should aim at
elimination of the infection from the flock. At present
the following procedures seem feasible:

1. The results obtained with flock No 1 of our field trial
   suggest that in a relatively short time a maedi-free
   flock can be obtained by artificial rearing of lambs
   which are separated from their dams immediately after
   birth. This procedure is based on the assumption that
   there is no intra-uterine transmission. It requires a
   strict hygienic regime which is difficult to perform
   under the usual farming conditions. The use of bovine
   colostrum could be considered to reduce neonatal infections. It seems to be the method of choice for experimental flocks. Recently by this approach maedi/visna
   virus was eliminated from a flock of the North Dakota
   State University (41).

2. In 1972, we initiated an attempt to develop a control
   programme on the basis of serological testing of a
   flock composed of 34 five year-old sheep and 41 of their
   two year-old progeny (flock No 3 of our field trial).
   The sheep were examined and sampled every six months and

the sera tested by AGPT and CFT. All clinically suspected
and all seropositive sheep were eliminated. No sheep were
introduced from outside, the ewes being served by rams
born in the preceeding year. Except for the closed management, the flock has been kept under average Dutch
farming conditions. The results of the serological tests
are presented in Table 3.

Table 3. Decline of number of seropositive sheep (AGPT +
   CFT) in a maedi infected flock by
   slaughter of serologically positive and clinically diseased animals at six month intervals.

	Sheep born in 1967	Sheep born in 1971 and later	Total percentage positives
1972	17/34	0/43	22.1
1973	9/17	5/41	24.1
1974	1/ 7	3/36	9.3
1975		0/52	0
1976		0/49	0
1977		0/84	0
1978		0/104	0

No further seropositive sheep were detected later than three
years after initiating this experiment. All 24 sheep which
have been slaughtered since then, were examined virologically
and by histological techniques. None of these sheep yielded
virus and no histological lung lesions were observed. We
tentatively conclude that maedi has been eradicated from
this flock. As mentioned earlier a number of virus infected
sheep do not react serologically and, following the above
procedure, will only be detected after clinical disease has
developed. The infection may thus linger in the flock for
several years without being traced. In order to define the
limits of usefulness of the above method the experiments
have been extended to a few other farms in the country. The
procedure of these field trials is as described above, but

in addition to slaughter of seropositive ewes, their offspring in also being eliminated since we observed a correlation in serological results between seropositive ewes and
their progeny. A regular culling of seropositive sheep is
certainly more economical than slaughter of all sheep of
infected farms, followed by restocking with maedi/visna-free
sheep. This procedure was successfully applied in Iceland
(24), but was only feasible as a desperate attempt to save
the sheep industry in this country.

REFERENCES

1. Sigurdsson, B: Rida, a chronic encephalitis of sheep with general remarks on infections which develop slowly and some of their special characeristics. Brit Vet J 110: 341-354, 1954.

2. Sigurdsson, B, H Thormar, PA Pálsson: Cultivation of visna virus in tissue culture. Arch ges Virusforsch 10: 368-381, 1960.

3. Sigurdardóttir, B, H Thormar: Isolation of a viral agent from the lungs of sheep affected with maedi. J Infect Dis 114: 55-60, 1964.

4. De Boer, GF: Zwoegerziekte. A persistent virus infection in sheep. Thesis, University of Utrecht: 1-211, 1970.

5. Ressang, AA, GF de Boer, GC de Wijn: The lungs in zwoegerziekte. Path Vet 5: 353-369, 1968.

6. Ressang, AA, FC Stam, GF de Boer: A meningo-leucoencephalomyelitis resembling visna in Dutch zwoeger sheep. Path Vet 3: 401-411, 1966.

7. Thormar, H, H Helgadóttir: A comparison of visna and maedi viruses II. Serological relationship. Res Vet Sci 6: 456-465, 1965.

8. Thormar, H: A study of maedi virus. Proc Int Conf on Lung Tumours. Perugia, Italy. Severi L (ed): 393-402, 1965.

9. De Boer GF: Zwoegerziekte virus, the causative agent for progressive interstitial pneumonia (maedi) and meningo-leucoencephalitis (visna) in sheep. Res Vet Sci 18: 15-25, 1975.

10. Harter, DH, R Axel, A Burny, S Gulati, J Schlom, S Spiegelman: The relationship of visna, maedi and RNA tumour viruses as studied by molecular hybridization. Virology 52: 287-291, 1973.

11. De Boer GF, C Terpstra, J Hendriks, DJ Houwers: Epizootiological studies of maedi/visna of sheep. Res Vet Sci, in press.

12. Sigurdsson, B, PA Pálsson, H Grímsson: Visna a demyelinating transmissible disease of sheep. J Neuropathol Exp Neurol 16: 389-403, 1957.

13. Georgsson, G, N Nathanson, PA Pálsson, G Pétursson: The
    pathology of visna and maedi in sheep. In: Slow virus
    diseases of animals and man, Kimberlin RH (ed) Amster-
    dam-Oxford, North Holland Publishing Company, 1976,
    p61-96.

14. Pálsson, PA: Maedi/visna, a slow virus disease. Bulletin
    OIE, in press.

15. Nathanson, N, H Panitch, PA Pálsson, G Pétursson,
    G Georgsson: Pathogenesis of visna. II. Effect of
    immunosuppression upon early central nervous system
    lesions. Lab Invest 35: 444-451, 1976.

16. De Boer, GF: Antibody formation in zwoegerziekte, a slow
    infection in sheep. J Immunol 104: 414-422, 1970.

17. Gudnadóttir, M, PA Pálsson: Host-virus interaction in
    visna infected sheep. J Immunol 95: 1116-1120, 1965.

18. Gudnadóttir, M, K Kristinsdóttir: Complementfixing
    antibodies in sera of sheep affected with visna and
    maedi. J Immunol 98: 663-667, 1967.

19. Terpstra, C, GF de Boer: Precipitating antibodies against
    maedi/visna virus in experimentally infected sheep.
    Arch ges Virusforsch 43: 53-62, 1973.

20. Haase, AT, O Stowring, D Narayan, DE Griffin, D Price:
    Slow persistent infection caused by visna virus: Role
    of host restriction. Science 195: 175-177, 1977.

21. Narayan, O, DE Griffin, J Chase: Antigenic shift of
    visna virus in persistently infected sheep. Science
    197: 376-379, 1977.

22. De Boer, GF, C Terpstra: The incidence of maedi/visna
    virus infections in the Netherlands. Tijdschr Dier-
    geneesk 99: 655-658, 1974.

23. Bruns, M, B Frenzel, F Weiland, OC Straub: Vergleich
    verschiedener serologischer Methoden zum Nachweis von
    Antikörpern gegen Maedi/Visna Virus. Zbl Vet Med B
    25: 437-443, 1978.

24. Pálsson, PA: Maedi and visna in sheep. In: Slow virus
    diseases of animals and man, Kimberlin RH (ed),
    Amsterdam-Oxford, North Holland Publishing Company,
    1976, p17-42.

25. Enchev, S: On the malignant (metastasizing) forms of
    what is known as pulmonary adenomatosis (jaagsiekte)
    in sheep. CR Acad Bulg Sci 16: 441-444, 1963.

26. Nobel, TA, F Neuman, U Klopfer: Metastases in pulmonary
    adenomatosis of sheep. Refuah Vet 25: 57-52, 1968.

27. Thormar, H, H von Magnus: Neutralization of visna virus
    by human sera. Acta Path Microbiol Scand 57: 261-267,
    1963.

28. Thormar, H: Visna-maedi infection in cell culture and in
    laboratory animals. In: Slow virus diseases of animals
    and man, Kimberlin RH (ed). Amsterdam-Oxford, North
    Holland Publishing Company, 1976, p97-114.

29. MacIntyre, EH, CJ Wintersgill, AE Vatter: A modification
    in the response of human astrocytes to visna virus.
    Am J Vet Res 35: 1161-1163, 1974.

30. Krogsrud, J, H Udnes: Maedi (progressive interstitial
    pneumonia in sheep). Diagnosis epizootiology, prevent-
    ion and control programme in Norway. Bulletin OIE, in
    press.

31. Straub, OC: Ueber die Isolierung von Maedi Visna Virus
    (MVV) aus einem deutschen Schafbestand. Berl Münch
    Tierärztl Wschr 83: 357-360, 1970.

32. Bruns, M, OC Straub, F Weiland, S Feiler: Detection of
    antibodies against glycoprotein of maedi/visna virus
    released as soluble antigen in cell cultures. In
    preparation.

33. Cutlip, RC, TA Jackson, GA Laird: Prevalence of ovine
    progressive pneumonia in a sampling of cull sheep from
    Western and Midwestern United States. Am J Vet Res 38:
    2091-2093, 1977.

34. Gates, NL, LD Winward, JR Gorham, DT Shen: A serologic
    survey of the prevalence of ovine progressive pneumonia
    in Idaho range sheep. In preparation.

35. Schaltenbrand, G, OC Straub: Visna/maedi in einer unter-
    fränkischen Merinofleischschafherde. Dtsch Tierärztl
    Wschr 79: 10-12, 1972.

36. Weinhold, E: Visna-Virus-ähnliche Partikel in der Kultur
    von Plexus Chorioideus Zellen einer Ziege mit Visna-
    Symptomen. Zbl Vet Med B 21: 32-36, 1974.

37. Weinhold, E, B Triemer: Visna bei der Ziege. Zbl Vet
    Med B 25: 525-538, 1978.

38. Dwivedi, JN: Studies on jaagsiekte and maedi (pulmonary
    adenomatosis complex) in sheep and goats with special
    reference to their infections vis-à-vis cancerous
    nature. Mathura, India; U.P. College of Vet Sci Husb:
    1-52, 1974.

39. Süveges, T, A Széky: Incidence of maedi (chronic progr-
    essive interstitial pneumonia) among sheep in Hungary.
    Acta Vet Acad Sci Hung 23: 205-217, 1973.

40. Sigurdsson, B, PA Pálsson, A Tryggvadóttir: Transmission
    experiments with maedi. J Infect Dis 93: 166-175,
    1953.

41. Molitor, TW, MR Light, IA Schipper: Elevated concentr-
    ations in serum immunoglobulins due to infection by
    ovine progressive pneumonia virus. Am J Vet Res 40:
    69-72, 1979.

DISCUSSIONS OF PAPERS BY G. PÉTURSSON ET AL. AND G. F. DE BOER AND D. J. HOUWERS

The possibility of studying the effect of immunosuppression in long-term experiments with visna was mentioned. Dr Pétursson replied that this was not possible as the animals succumbed rapidly; possibly a modified immunosuppressive regime might be feasible. The presence of antibodies to visna in the cerebrospinal fluid in very high titre was queried; these antibodies were indeed present, sometimes in titres up to four times greater than those observed in the peripheral blood.

Antigenic variation in visna infections was discussed. This was a real phenomenon and variants of virus were recovered at different times from the same animal, the new variant being less readily neutralised by early serum than the early variant. Much more systematic work was necessary in this area.

THE KEY ROLE OF CELL MEMBRANE MODULATION IN THE BIOLOGICAL EFFECTS OF INTERFERON

C. CHANY, M. F. BOURGEADE, M. BERGERET, D. SERGIESCU, A. PAULOIN AND
F. CHANY-FOURNIER

Interferon has been regarded for a long time only as an inducible antiviral protein released by virus-infected cells. However, a great variety of antigens of bacterial origin (1), lectins (2), or other substances, can induce interferon in animals or white blood suspensions. This observation greatly complicates the relatively simple original concept on the role of these substances in acute or chronic infections. In the present state of knowledge interferons can be divided into two main cat gories:

- Type I interferon is produced in virus-infected B lymphocytes (3) or other somatic cells. Although leukocyte interferon is different from the fibroblast variant in molecular weight and antigenic structure (4-6), they are all pH resistant and heat-sensitive.

- Type II interferon is released by antigen-sensitized T lymphocytes after restimulation by the same antigen or lectins. In contrast, this variant is heat-resistant, but acid-labile (Table 1).

Table 1

Interferon	Lymphocytes	Other Somatic Cells	Inducer
Type I	B	+	Virus RNA
Type II	T	0	Viral or other antigens (DH)[x]

[x]DH means delayed hypersensitivity

Many aspects of type I interferon production and action are well documented (7). Not much is known, however, on the biological role of type II interferon. We will, therefore, focus on presently available knowledge on type I interferon and point out possible common features of the two types.

During acute infection, interferon is produced locally by the infected cells and by both B and T lymphocytes. It diffuses thereafter into the inflammatory area and into the bloodstream (8). It acts on target cells which can be located at distant sites. In this respect interferon resembles other messenger molecules such as lymphokines or hormones. The effect on target cells is a general slow-down of cell metabolism which results in vitro in a decrease of cell replication (9) and is associated with the development of the antiviral state. It is as yet uncertain whether the antiviral activity appears or not in interferon-treated cells in the absence of a challenge virus (10). Studies of interferon action in cell-free systems seem to indicate that dsRNA has to be added to the cell sap to induce or increase a number of metabolic steps which result in the inhibition of viral protein synthesis and the degradation of mRNA (11-13). There is no evidence whatsoever that these inhibitory steps are really selective for virus metabolism. If this is correct, it means that the challenge virus triggers or enhances the antiviral state in interferon-treated cells. Thus in the absence of infection, interferon could modify somewhat differently the biological status of the cell.

The primary site of the biological action of interferon is a cell membrane modulation followed, as first demonstrated by Paucker (9), by a general repressing effect on cell metabolism. In addition, a change in the phenotypic expression occurs. This can be shown when cancer cells are carried for a long period in the presence of interferon. These cells recover contact inhibition and no longer produce colonies in soft agar, even though the cells continue to divide at a lower rate (14, 15) (Fig. 1).

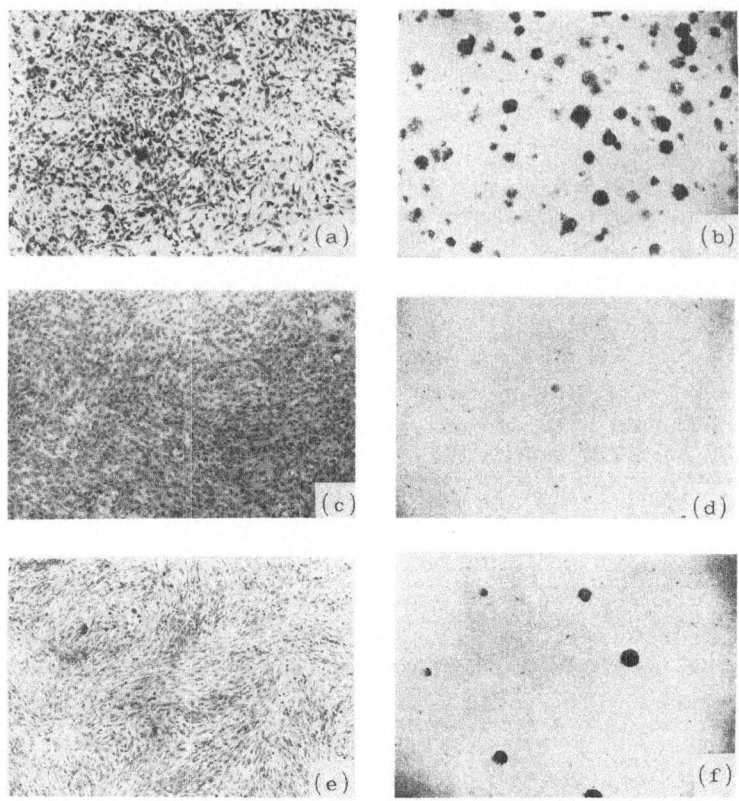

Figure 1[x]

(a) Murine sarcoma virus (MSV)-transformed mouse embryonic
     fibroblasts (MEF).

(b) Same cells producing colonies in soft agar.

(c) Same cells grown in the presence of interferon for
     200 passages (MSV-IF+).

(d) MSV-IF+ cells grown in soft agar. Practically no
     development of colonies.

(e) Normal MEF used for transformation by MSV.

(f) Reappearance of colonies in MSV-IF+ cells when inter-
     feron is omitted from the tissue culture medium for
     30 passages.

[x]Reprinted from the J. Gen Virol. (1970), _7_, 206-207 with
slight modification.

Thus, the mechanism of cell membrane modulation is, in our view, the key problem for a better understanding of the apparently contradictory effects which can be produced by interferon during chronic infection or disease.

ROLE AND MECHANISM OF CELL MEMBRANE MODULATION IN THE DEVELOPMENT AND MAINTENANCE OF THE ANTIVIRAL STATE

Binding of interferon to the cell. It has been shown that in order to act interferon has to bind to the cell membrane (16). Binding is a temperature-independent step and can occur at 4°C, but activation of the antiviral state only appears at 37°C. Further studies in somatic Cercopithecus monkey-mouse (or human-mouse) cells have shown that the presence of only one monkey 29 chromosome (or human 21) in every cell is necessary for the expression of primate interferon sensitivity. Trypsin treatment of one monkey-mouse hybrid clone, followed by the suspension of the cells in a serum-free medium for 3-4 h, abolishes the sensitivity of cells to primate, but not to mouse, interferon (Table 2). This experiment suggests that the primate interferon receptor, more exposed than the murine variant, is temporarily destroyed by the proteolytic enzyme. It can be concluded therefrom that a membrane-bound glycoprotein is responsible for interferon binding and is probably the interferon cell species specific component. It is also likely that in the hybrid cells, the mouse components are responsible for the subsequent cytoplasmic changes which occur in the cell (17).

The existence of another but non-specific binding site has been suggested by Besançon and Ankel (18). Interferon covalently bound to sepharose can induce antiviral activity in sensitive cells after contact in the absence of detectable leakage (19, 20). Such sepharose-bound interferon loses its antiviral properties when gangliosides (in the case of murine interferon GM 2) (21) or lectins (PHA) (22) are added to the beads. The role of gangliosides in interferon action is also substantiated by (a) interferon resistance of cells lacking mature monosialogangliosides (23) and (b) disappearance of interferon sensitivity induced by cho-

Table 2[x]

		Interferon			
		Human WBC		Mouse	
Cell type	Dispersing agent	No serum	2% calf serum	No serum	2% calf serum
Parental monkey	Trypsin	0.07	3.42	N.D.	N.D.
Hybrid clone M	Trypsin	0.49	2.06	2.76	2.86
Parental mouse	Trypsin	N.D.	N.D.	1.13	1.75
Mouse L cells	Trypsin	N.D.	N.D.	2.80	2.97
Parental monkey	EDTA	2.86	3.07	N.D.	N.D.
Hybrid clone M	EDTA	2.23	2.25	N.D.	N.D.

Antiviral effect of primate and mouse interferon in trypsinized or EDTA-dispersed parental and clone cells. Numbers represent $\log_{10}$ inhibition of VSV yield in interferon-treated cells as compared to contact cells subjected to the same experimental conditions but incubated with media containing no interferon. N.D. means not done.

[x]Reprinted from Proc. Nat. Acad. Sci. USA 70 (1973) 559, with some modifications.

---

lera toxin or hormones which can also bind to gangliosides (24), although the effect is probably indirect.

Cell membrane modulation. As here shown, interferon binds to at least two cell membrane constituents. On the other hand, the now well-established model of membrane structure, devised by Singer and Nicolson (25), predicts the lateral mobility of the glycoproteins and glycolipids which are its structural parts. The sigmoidal shape of the interferon dose-response relationship can be especially well analyzed in somatic primate-mouse hybrid cells which, as has been shown earlier, contain at least two interferon species-specific receptor proteins. Indeed, in the same hybrid cell, the dose-response relationship can be different for mouse or primate interferon. It is therefore likely that this modification is due to the number, exposure, or activity of these receptors. The expected results will be a remodulation of the cell membrane depending on the concentration of the interferon molecules and on the mobility of the membrane constituents.

Membrane-bound glycoproteins are anchored on cytoplasmic cytoskeletal structures (thin filaments made up of actin, thick filaments formed by myosin in a hammer-like structure, and tubuline). These constituents govern the lateral mobility of the cell membrane-bound glycoprotein. Their integrity could be important for interferon action.

Indeed, data available show that inhibition of the cytoskeletal system by drugs such as cytochalasin B, vinblastine, or colchicine inhibit in parallel the antiviral action of both type I and II interferons (26, 27). In cells infected with Sarcoma virus temperature-sensitive in Sarcoma (src) gene, the shift from non-permissive to permissive temperature results in the loss of the synthesis of microtubules and microfilaments (28, 29). Sarcoma cells containing defective cytoskeletons also respond poorly to interferon. Sodium butyrate, which improves in such cells the synthesis of these components, also enhances interferon sensitivity, while the same drug is inactive in normal cells (30, 31).

The interferon-induced modifications of the cell membrane are not only necessary to establish the antiviral (or other) effects of interferon, but are also necessary for its maintenance. When this interferon-induced distribution of cell membrane constituents is modified either from the inside (by inhibitors of the cytoskeleton) or from the outside by PHA (32), the established antiviral state decays in 6-8 h and the cell recovers sensitivity to viruses. When the cells are retreated with interferon, antiviral activity can be reinduced.

RELATIONSHIP OF CELL MEMBRANE MODULATION **TO** NON-ANTIVIRAL
EFFECTS OF INTERFERON

Effects on the immune system. In addition to its
effect on the reduction of cell proliferation and changes
in the phenotypic expression of the cells already mentioned, interferon also has a complex effect on the immune
system. Phagocytosis of carbon particles by mouse peri-
toneal macrophages can be increased (33). This can explain
the suppression of infection due to organisms such as
E. Coli, Listeria monocytogenes, Chlamidia, plasmodia
Berghei, etc.

T-lymphocytes sensitized to L 1210 cells incubated in
the presence of interferon show enhanced cell mediated cytotoxicity (34). More complex results have been obtained
using T cells sensitized to a monkey-mouse hybrid cell
after the immunization of syngeneic mice. Interferon
treatment of the target cells increases cytotoxicity after
primary (low level) sensitization. On the contrary, the
same treatment protects target cells against lymphocytes
hypersensitized after secondary in vitro treatment with
the target cells. It is not difficult to relate these
contradictory effects to cell membrane modulation, since
interferon-treated target cells show, in addition, a significantly increased spontaneous lysis in the presence of
^{51}CR. Such surface changes could be responsible for the
increased expression of H-2 antigens on interferon-treated
lymphocytes (35).

Interferon has been shown to act on delayed hypersensitivity in mice. Using the ear-swelling and footpad-
swelling tests, interferon decreases or completely inhibits
sensitization of mice when injected prior to the antigen or
decreases its expression in already sensitized mice (36).

B cell functions can also be affected by interferon.
Small amounts of interferon can increase, while large amounts
decrease, antibody production in vitro (37). In addition,
interferon treatment of T lymphocytes inhibits lymphoblastoid
transformation by lectins (38).

Effects of the two interferons during acute and chronic infections. As stated earlier, the problem is relatively simple during primary infection. In the inflammatory area, type I interferon is produced and after hypersensitization of the T cells in a later step, type II interferon can also be secreted. Indeed, it has been shown that sensitized T lymphocytes can produce type II interferon in the presence of macrophages when triggered by Vaccinia (39) or Herpes (40) viruses. Interferon, after induction of the here described cell membrane modification, will induce a number of metabolic steps leading to the blockage of virus infection. An important link between cell membrane and cytoplasmic events is missing. The possible role of cyclic nucleotides is uncertain (41).

In chronic infections the problem is much more complex. Undoubtedly type I and II interferons will similarly protect uninfected cells and limit the extension of the lesions.

In cells already infected by the virus, several possibilities have to be considered. The virus replicates in the cells without producing a lethal injury. For RNA tumor viruses it has been shown that interferon can limit virus replication to some extent, acting much less on viral protein synthesis than on the budding of the particles from the cell membrane. They could become not only reduced in quantity but also altered in quality, resulting in the production of non-infectious particles (42-44).

In other cases, cells can be modified during an incomplete viral cycle. This is true for SSPE where nothing is known at present about the sensitivity of such cells to interferon. Several therapeutic trials using purified human interferon, injected either by the subcutaneous or intrathecal route, fail to show any modification in the course of the disease (unpublished data). Similar results have been observed during the acute phase of rabies. During this infection, patients can be maintained alive for about three weeks and die very often when the circulating antibody rises to a peak level. Interferon inhibits to a certain extent antibody synthesis but is without effect on the clinical expression of the disease (unpublished data).

However, during chronic hepatitis B, where the infectious virus is expressed, intensive and prolonged interferon treatment has at least in some cases beneficial effects. A few patients have been completely cleared of the virus, with disappearance of all biological manifestations of the disease (45).

Interferon as an aggravating factor in chronic disease. In two experimental diseases interferon seems to be an aggravating factor in the pathogenic course of the disease. During choriomeningitis infection in mice, injection of antibody to mouse interferon protects mice against glomerulonephritis by neutralizing the endogenous interferon induced by the virus. This protection occurs in spite of a significant increase of virus titer in the animals (46).

NZB mice and their NZB/NZW hybrids develop an auto-immune disease comparable to human Systemic Lupus Erythematosus. In NZB mice, the main character of the disease is the appearance of anti-red cell antibodies and hemolytic anaemia. In NZB/NZW hybrids, anti-nuclear factors and anti-DNA antibodies are found associated to immune complex resulting in glomerulonephritis which eventually kills the animal. The administration of interferon inducers (47-49) or prolonged interferon treatment starting at birth (50) increases mortality in the animals when compared to the placebo group. It is of interest that in interferon-treated NZB mice, prior to death, anti-nuclear antibodies can be detected although they are generally only present in the NZB/NZW hybrids. Most of the animals develop a reticulum cell sarcoma which could contribute to their premature death. It is unknown by what mechanism interferon produces such unusual adverse effects. It is likely that modifications of cellular immunity (such as accelerated decrease of repressor T cell functions, already deficient in these animals) could be involved.

In summary, the biological role of interferon is more complex than its antiviral action. It is possible that type II interferon could regulate immune functions especially during delayed hypersensitivity. However, not much is known about the metabolic processes it induces in the cell. It is likely that some steps are common (51) with type I, while other are different (27).

Type I interferon is produced by practically all somatic cells during viral, or some bacterial infections. In both cases its synthesis and different biological effects are submitted to a negative control mechanism. Thus both production and action are transitory phenomena. Cell membrane modulation seems to be the key step for the induction, primary amplification, maintenance, and degradation of most, if not all, biological effects (52).

REFERENCES

1. Youngner, JS, SB Salvin: Production and properties of migration inhibiting factor and interferon in the circulation of mice with delayed hypersensitivity. J Immunol 14: 1914-1922, 1973.

2. Wheelock, EF: Interferon-like virus inhibitor induced in human leukocytes by phytohemagglutinin. Science 149: 310-311, 1965.

3. Yamaguchi, T, K Handa, Y Shimizu, T Abo, K Kumagai: Target cells for interferon production in human leukocytes stimulated by Sendaï virus. J Immunol 118: 1931-1935, 1977.

4. Duc-Goiran, P, B Galliot, C Chany: Studies on virus-induced interferons produced by the human amniotic membrane and white blood cells. Arch Virusforsch 34: 232-243, 1971.

5. Levy-Koenig, RE, MJ Mundy, K Paucker: Immunology of interferons.I. Immune response to protective and non-protective interferons. J Immunol 104: 785-790, 1970.

6. Havell, EA, B Berman, CA Ogburn, K Berg, K Paucker, J Vilcek: Two antigenically distinct species of human interferon. Proc Nat Acad Sci USA 72: 2185-2187, 1975.

7. Baron, S, F.Dianzani (eds): Texas reports on biology and medicine. The interferon system: A current review to 1978, 35, University of Texas Medical Branch, 1977.

8. Baron, S: Mechanism of recovery from viral infection. In Advances in virus research 10, Smith,DA, MA Lauffer (eds), Academic Press Inc., New York, 1963, p.39-60.

9. Paucker, K, K Cantell, W Henle: Quantitative studies on viral interference in suspended L cells. III. Effect of interfering viruses and interferon on the growth rate of cells. Virology 17: 324-334, 1962.

10. Friedman, RM, DH Metz, RM Esteban, DR Tovell, LA Ball, IM Kerr: Mechanism of interferon action: inhibition of viral messenger ribonucleic acid translation in L-cell extracts. J Virol 10: 1184-1198, 1972.

11. Kerr, IM, RE Brown, LA Ball: Increased sensitivity of cell-free protein synthesis to double-stranded RNA after interferon treatment. Nature (London) 250: 57-59, 1974.

12. Farrell, PJ, GC Sen, MF Dubois, L Ratner, E Slattery, P Lengyel: Interferon action: two distinct pathways for inhibition of protein synthesis by double-stranded RNA. Proc Nat Acad Sci USA 75: 5893-5897, 1978.

13. Revel, M: Interferon-induced translational regulation. In Texas reports on biology and medicine, Baron, S, F Dianzani (eds), 35, 1977, p. 212-218.

14. Chany, C, M Vignal: Etude du mécanisme de l'état réfractaire des cellules à la production d'interféron, après inductions répétées. Compt Rend Acad Sci Paris 267: 1798-1800, 1968.
15. Chany, C, M Vignal: Effect of prolonged interferon treatment on mouse embryonic fibroblasts transformed by murine sarcoma virus. J Gen Virol 7: 203-210, 1970.
16. Friedman, RM: Interferon binding: the first step in establishment of antiviral activity. Science 156: 1760-1761, 1967.
17. Chany, C, A Grégoire, M Vignal, J Lemaître-Moncuit, P Brown, F Besançon, H Suarez, R Cassingena: Mechanism of interferon uptake in parental and somatic monkey-mouse hybrid cells. Proc Nat Acad Sci USA 70: 557-561, 1973.
18. Besançon F, H Ankel: Binding of interferon to gangliosides. Nature 252: 478-480, 1974.
19. Ankel, H, C Chany, B Galliot, MJ Chevalier, M Robert: Antiviral effect of interferon covalently bound to sepharose. Proc Nat Acad Sci USA 70: 2360-2363, 1973.
20. Chany, C, H Ankel, B Galliot, MJ Chevalier, A Grégoire: Mode of action and biological properties of insoluble interferon. Proc Soc Exp Biol Med 147: 293-299, 1974.
21. Besançon, F, H Ankel, S Basu: Specificity and reversibility of interferon ganglioside interaction. Nature 259: 576-578, 1976.
22. Besançon, F, H Ankel: Inhibition of interferon action by plant lectins. Nature 250: 784-786, 1974.
23. Vengris, VE, FH Reynolds, Jr, MD Hollenberg, P Pitha: Interferon action: role of membrane gangliosides. Virology 72: 486-493, 1976.
24. Friedman, RM, LD Kohn: Cholera toxin inhibits interferon action. Biochem Biophys Res Commun 70: 1078-1084, 1976.
25. Singer, SJ, G Nicolson: The fluid mosaic of the structure of cell membranes. Science 175:720-731, 1972.
26. Bourgeade, MF, C Chany: Inhibition of interferon action by cytochalasin B, colchicine and vinblastine. Proc Soc Exp Biol Med 153: 501-504, 1976.
27. Bourgeade, MF, C Chany, TC Merigan: Type I and type II interferons: differential antiviral actions in transformed cells. J Gen Virol, submitted for publication.
28. Wang, E, AR Goldberg: Effects of the src gene product on microfilament and microtubule organization in avian and mammalian cells infected with the same temperature sensitive mutant of Rous Sarcoma Virus. Virology 92: 201-210, 1979.
29. Pollack, R, M Osborn, K Weber: Pattern of organization of actin and myosin in normal and transformed cultured cells. Proc Nat Acad Sci USA 72: 994-998, 1975.
30. Bourgeade, MF: Effet du butyrate de sodium sur la sensibilité à l'interféron de cellules normales et transformées par le virus du sarcome murin. Compt Rend Acad Sci Paris 187: 391-394, 1978.

31. Bourgeade, MF, C Chany: Effect of sodium butyrate on the antiviral and anticellular action of interferon in normal and MSV transformed cells. Submitted for publication.

32. Pauloin, A, F Chany-Fournier, L Epstein, C Chany: Dégradation par la phytohémagglutinine de l'état antiviral induit par l'interféron. Compt Rend Acad Sci Paris 284: 1119-1122, 1977.

33. Huang, K: Effect of interferon on phagocytosis, in Texas reports on biology and medicine, Baron, S, F Dianzani (eds), 35, 1977, p. 350-356.

34. Lindhal, P, P Leary, I Gresser: Enhancement by interferon of the specific cytotoxicity of sensitized lymphocytes. Proc Nat Acad Sci USA 69: 721-725, 1972.

35. Bergeret, M, A Grégoire, C Chany: Different effect of interferon on target cells after treatment with lymphocytes exposed to primary or secondary sensitization. Submitted for publication.

36. De Maeyer-Guignard, J, A Cachard, E De Maeyer: Delayed-type hypersensitivity to sheep red blood cells: inhibition of sensitization by interferon. Science 190: 574-576, 1975.

37. Gisler, RM, P Lindhal, I Gresser: Effects of interferon on antibody synthesis in vitro. J Immunol 113: 438-444, 1974.

38. Lindhal-Magnusson, P, P Leary, I Gresser: Interferon inhibits DNA synthesis induced in mouse lymphocyte suspensions by phytohaemagglutinin or by alloqeneic cells. Nature New Biol 237, 120-121, 1972.

39. Epstein, LB, DA Stevens, TC Merigan: Selective increase in lymphocyte interferon response to vaccinia antigen after revaccination. Proc Nat Acad Sci USA 69: 2632-2636, 1972.

40. Rasmussen, LE, GW Jordan, DA Stevens, TC Merigan: Lymphocyte interferon production and transformation after herpes simplex infections in humans. J Immunol 112: 728-735, 1974.

41. Friedman, RM, I Pastan: Interferon and cyclic-3'5'-AMP: potentiation of antiviral action. Biochem Biophys Res Commun 36: 735-740, 1969.

42. Pitha, PM, WP Rowe, MN Oxman: Effect of interferon on exogenous, endogenous, and chronic murine leukemia virus infection. Virology 70: 324-338, 1976.

43. Friedman, RM, JM Ramseur: Inhibition of murine leukemia virus production in chronically infected AKR cells; a novel effect of interferon. Proc Nat Acad Sci USA 71: 3542-3544, 1974.

44. Billiau, A, H Sobis, P De Sommer: Influence of interferon on virus particle formation in different oncornavirus carrier cell lines. Int J Cancer 12: 646-653, 1973.

45. Greenberg, HB, RB Pollard, LI Lutwick, et al: Effect of human leukocyte interferon on hepatitis B virus infection in patients with chronic active hepatitis. New Eng J Med 25: 517-522, 1976.

46. Gresser, I, L Morel-Maroger, P Verroust, Y Rivière, JC Guillon: Anti-interferon globulin inhibits the development of glomerulonephritis in mice infected at birth with lymphocytic choriomeningitis virus. Proc Nat Acad Sci USA 75: 3413-3416, 1978.

47. Steinberg, AD, S Baron, N Talal: The pathogenesis of auto-immunity in New Zealand mice. I. Induction of antinucleic acid antibodies by polyinosinic-polycytidylic acid. Proc Nat Acad Sci USA 63: 1102-1107, 1969.

48. Walker, SE: Accelerated mortality in young NZB/NZW mice treated with the interferon inducer, tilorone hydrochloride. Clin Immunol Immunopathol 8: 204-212, 1977.

49. Sergiescu, D, I Cerutti, E Efthymiou, A Kahan, C Chany: Adverse effects of interferon treatment on the life span of NZB mice. Biomedicine, in press.

50. Heremans, H, A Billiau, A Colombatti, J Hilgers, P De Somer: Interferon treatment of NZB mice: accelerated progression of autoimmune disease. Infect Immun 21: 925-930, 1978.

51. Wietzerbin, J, S Stephanos, R Falcoff, E Falcoff: Some properties of interferons induced by stimulants of T and B lymphocytes. In Texas reports on biology and medicine, Baron, S, F Dianzani (eds), 35, 1977, p. 205-211.

52. Chany, C: Membrane-bound interferon specific cell receptor system: role in the establishment and amplification of the antiviral state, Biomedicine 24: 148-157, 1976.

DISCUSSION OF PAPER BY C. CHANY

This discussion opened with a question as to whether conconavelin A was more effective than phytohaemagglutinin in blocking receptors for interferon. Dr Chany replied that there was no evidence for this. There was one report that the receptors could be 'capped' but this could not be accepted. Gangliosides gc2 and gc3 were the most important glycoproteins concerned with the action of interferon; a relevant observation in this context was that transformed cells were less sensitive to interferon than non-transformed cells.

Dr Lachmann asked for further information on interferon types I and II. In reply, Dr Chany outlined the facts as known. Type I is produced by leucocytes and also by other cells (for example, fibroblasts). It is pH resistant glycoprotein. Type II interferon was produced by T lymphocytes, following previous sensitisation by antigens and was a heat-stable, pH-labile glycoprotein. The actions of types I and II interferon differed, type II interferon having a greater anti-cellular and a lesser anti-viral effect than type I interferon. Types I and II interferon were antigenically different and there was not complete cross-neutralisation.

The mechanism of action of type II interferon was not known, although it appeared that the two types shared receptors. An interesting observation was that trisomy-21 cells were more susceptible to both types of interferon than normal diploid cells.

IMMUNE RESPONSES IN THE CEREBROSPINAL FLUID

A. LOWENTHAL AND D. KARCHER

The immune response may be either referred to cellular or to humoral immunity in cerebrospinal fluid (CSF) as in all other biological fluids. The study of cellular immunity in CSF has never been thoroughly undertaken, however, recent work has brought important information in this field, and allowed to identify the T lymphocytes (1).

The study of humoral immunity, on the other hand is much more important and a large literature is devoted to this subject. These researches enabled us to discover with agar electrophoresis the phenomenon of 'restricted heterogeneity' in human CSF (2). Different other electrophoretic methods and quantitative measurements, mainly immunological methods were later utilized and confirmed our results. In addition, culture of CSF cells showed that these cells produce immunoglobulins (3, 4). These points will be discussed in relation to slow viral diseases such as multiple sclerosis (MS), subacute sclerosing panencephalitis (SSPE), visna and distemper. On the other hand, we shall only mention the spongiform encephalopathies where all the examinations gave normal results and experimental allergic encephalomyelitis (EAE) where information on CSF is still incomplete.

MATERIAL AND METHODS.

A. Patients. For patients, and specially those affected with MS, the first need which arises is a clear clinical definition of the disease. A rigorous definition has to be reached before starting to study the biological material and the CSF. A clinical case of multiple sclerosis cannot, in our opinion, be accepted for scientific investigations without a prior CSF electrophoretic examination. Cases with increased cell counts in the CSF, frequently used by some authors, should be considered as exceptional.

Besides MS, SSPE, other acute encephalitis, such as necrotizing encephalitis, and chronic encephalitis, such as trypanosomiasis, filariasis, neurosyphilis as well as acute and chronic, bacterial and viral meningitis, can provide material for investigation. This material may either be of human or of animal origin.

B. Identification of <u>T lymphocytes</u> can be done by methods described by Kam-Hansen (1). For the cell cultures, we will mention in particular Sandberg-Wollheim's work (3,4).

C. Identification and determination of <u>immunoglobulins</u>.

    1) electrophoretic methods. The electrophoretic examinations are practically always effected on concentrated CSF. The question arises whether concentration produces artefacts and therefore method applicable to the study of non-concentrated CSF seems indicated. Besides these methods, there are some indirect methods such as colloïdal reactions, formation of double ring precipitations and immunoelectrophoresis. In our opinion, their contribution to the study of the immunological phenomena observed in CSF is lesser than the electrophoretic methods.

    2) quantitative determinations of the immunoglobulins, IgG as well as IgM, were often carried out. They allow either a direct interpretation of the results or an interpretation based on the calculation of ratios which allows comparions of the distribution of immunoglobulins and other proteins (albumin e.g.) in CSF and in serum. We mention this approach although we give preference to direct quantitative determination. The restricted heterogeneity of the IgG's as seen in MS or SSPE makes, in our opinion a comparison with the so called homogeneous albumin fractions very difficult.

3) information relative to the immune reaction in CSF can also be obtained by measuring free or bound κ and λ light chains. The results confirm those of protein electrophoresis.

RESULTS

a) research on cellular immunity suggests that there is a decrease of the T cell depressive activity in CSF (1). In 30 % of cases of multiple sclerosis the number of cells is increased. These cells are usually lymphocytes, and in particular T lymphocytes (95 %).

The depressive immune activity of these T cells could be decreased, or perhaps, according to some authors their stimulating activity increased. The study of cellular immunity in the blood of patients suffering from multiple sclerosis has not provided so far any difenite information. The same is true for numerous slow viral diseases, with the exception perhaps of SSPE. In summary, in CSF, we are faced with an undoubted and important phenomenon : the modification of T lymphocytes.

b) studies investigating humoral immunity showed in human CSF, (before it was described in experimental animals) that there may be γ globulin fractionation, later called oligoclonal reaction (fig. 1). We think that the title 'antibody of restricted heterogeneity' used in experimental work in recent years, is more appropriate. It is probable that the detection of this reaction was more easy to achieve in CSF than in serum for purely technical reasons : the fact that in normal CSF there are practically no γ globulins. It has to be underlined that this observation was made first in CSF and thus is a neurological contribution. It was shown later that these γ globulins are mainly IgG's and only recently some papers referred to IgM's. This fractionation of the IgG is not specific for a disease. Once the fractionation

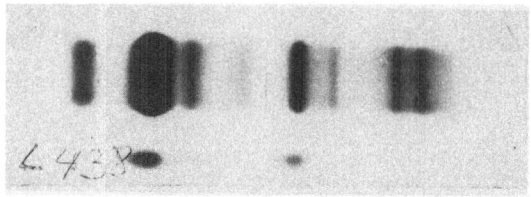

Figure 1 : Cerebrospinal fluid protein electrophoresis
with fractionation of the γ globulins.

has settled in, it remains qualitatively the same for
years. The quantitative interpretation of the phero-
grams is not simple; for this reason it is not easy
to demonstrate that this fractionation remains quantitatively constant. It persists, even after a complete
and protracted clinical recovery from some diseases -
for instance necrotizing encephalitis. According to
Vandvik, these fractionated immunoglobulins would be
Igl's (5).

c) The IgM's (6, 7, 8, 9, 10, 11, 12) are frequently, but
not always, increased in the CSF of patients affected
with multiple sclerosis. The fractionation of the IgG
and an increase of the IgM (11) are very often associated. Increases of the IgM may reach up to 5 times
their normal values. There are still technical problems
in the interpretation of the IgM determinations. The
IgA (7, 8, 12) may also be increased. The concentrations of IgA, IgM and IgG can vary independently.

DISCUSSION

We wish to discuss four different points :

A. What definition can we presently give to the immune
reaction, observed in the CSF in slow viral diseases
as SSPE and MS ?

B. Should the IgG, present in the CSF in cases of slow viral
diseases, be considered as homogeneous, like the IgG in
experimental hyperimmunization ? What are the known
immune activities of these homogeneous IgGs ?

C. where does the synthesis of these IgG's take place ? In
the CSF or in other organs or tissues ?

D. what is the physiopathological significance of this
hyperimmune process ? Is it a physiological or a
pathological reaction ?

A. In MS we could define the immune reaction, observed in
CSF, as characterized by a decrease of the depressive
activity of the T lymphocytes, associated with a restricted heterogeneity of the IgG and occasionally an increase of the IgM. Due to the fact that immunoglobulins
with restricted heterogeneity are technically easier to
show in CSF than in serum, one came to the conclusion,
maybe too hastily, that these immunoglobulins are produced locally in the central nervous system.

   Moreover, in 80 % of cases of SSPE, in some cases of MS
($\pm$ 30 %) and in diseases such as the Guillain-Barré syndrome, the fractionation is observed in serum as well
as in CSF. This is also the case in acute encephalitis,
like necrotizing encephalitis and in some chronic encephalitis, such as trypanosomiasis and filariasis. The
restricted heterogeneity of the immunoglobulins could
also be shown indirectly in MS serum, (it is observed
only sporadically by electrophoretic methods)
(fig 2) by precipitating the γ globulins with several
antigens (13).

Figure 2 : serum protein electrophoresis in multiple scle-
          rosis with fractionation of the γ globulins.

It seems probable that restricted heterogeneity is a general phenomenon that can be found in CSF and in serum in many neurological (and non neurological) diseases. Restricted heterogeneity in serum is a typical consequence of experimental hyperimmunization.

Until now, no investigation has been undertaken in the field of experimental hyperimmunization, to find out if oligoclonal reactions seen in the serum can also be found in CSF. Yet we can assume that they will probably be seen in CSF and this for two reasons :

1) the same homogeneous γ globulins are observed in serum and in CSF, in cases of myeloma, which are undoubtedly non-neurological diseases and where the homogeneous immunoglobulins have a non-neurological origin,

2) in ataxia telangiectasia, a disease characterized by a deficiency of IgA, a restricted heterogeneity of the IgG appears in serum, after repeated infections or after parenteral treatment with γ globulins (14). Undoubtedly, it is difficult to admit that these IgG are produced in the central nervous system.

Thus it is not excluded, and even highly probable, that a fractionation of immunoglobulins due to extra-neural processes will also be reflected in the CSF. Nor is it excluded that, in diseases where no anomalies in the serum are observed, as is frequently the case in MS, the anomalies in the serum may reamin hidden for technical reasons. We can thus assume that fractionation of the γ globulins can be considered a general phenomenon which probably develops to a maximum in the CSF, in disease states.

As far as we are concerned, we would consequently define the immune reaction observed in diseases such as SSPE and even MS, as a general hyperimmunization with probably an inhibition of the depressive action of the T lymphocytes. The hypothesis of hyperimmunization is confirmed by the fact that the measles antibody levels in SSPE are extremely high in serum and tissues, as well as by the presence of antibody of restricted heterogeneity. Quantitatively, with regard to the other proteins, the restricted heterogeneity is also more intense in CSF than in serum. In MS we believe that the humoral immune reaction is similar to that in SSPE.

B. Experimental hyperimmunization in animals leads to the formation of partially specific homogeneous immunoglobulins. The number of fractions experimentally obtained varies from case to case. Very often 2 main and 5 secondary fractions are found. In addition to these fractions, CSF and especially serum display a background of heterogeneous immunoglobulins. Similar electropherograms are observed in serum of patients suffering from MS or SSPE. The multiplicity of these fractions raises some questions : are they different antibodies, active against different antigenic sites of single protein, complex proteins, or are each of these fractions active against different viral antigens, or are they immun-complexes? The measles virus consists of numerous proteins, different from one another, and could thus stimulate different antibodies. This could be one explanation for the fractions multiplicity in SSPE; it is still, at present, only an hypothesis of some authors who showed that different viral proteins can precipitate various IgG .

Some comments ought to be made. It has to be noted on the one hand, that precipitation does not involve all the serum fractions, and on the other hand, that the CSF IgG, at least in MS and to a lesser degree in SSPE (13), is immunologically active against numerous and various viral antigens. It is difficult in these experiments to state precisely what is to be referred to the homogeneous fractions and what to the background in the Y globulin region after agar gel electrophoresis. Only precise quantitative determinations could answer this question. The question can also be raised whether hyper-

immunization provokes a specific or a generalized stimulation of the humoral immune response. The increase of
antibody titres against the antigen in experimental
hyperimmunization and the measles antibody titres in
SSPE, argues in favour of specificity. Specificity however
cannot be considered to have been demonstrated, as long as
quantitative measurements have not been carried out.
We shall return to this in connection with SSPE serum.

Hence two questions are raised with regard to the
fractions seen in neurological diseases : are they
homogeneous and/or are they specific? We have been
able to demonstrate in SSPE that the cathodic IgG serum
fraction after electrophoresis is homogeneous by the
study of the terminal aminoacid sequence of the variable
part of the light and the heavy chains (15).

Confirmation of this homogeneity is given by the
calculation of the κ and λ light chains'ratios. These
ratios reveal a predominance either of the κ or the λ
chains in serum and in CSF of SSPE and MS. The answer
is yet not as clear as in myelomatosis where homogeneity
has been proven beyond doubt.

What are these homogeneous immunoglobulins? Attempts
were made to identify them by direct methods or by absorption methods. Fractionation of the immunoglobulins
on Sephadex or DEAE columns or by isoelectric focusing
and thereafter measurement of their measles antibody
activity show very clearly in the case of SSPE that
only some fractions, and mainly the most cathodic ones,
have a very high measles antibody activity (16). The
curve obtained by measuring the measles antibody activity in the various SSPE serum IgG fractions after isoelectric focusing shows two main, very cathodic peaks,
and runs parallel with the optical density curve measured
for the same proteins after agar gel electrophoresis.

This is only partial evidence that the γ globulins are
measles antibodies. In absorbing this immunoglobulins
with antigens, in the case of SSPE with measles virus,

it is possible to precipitate them out partially if not
totally. Other antigens, different from the measles virus can, in MS, and even in SSPE, partially precipitate
out some IgG. Precipitation is qualitatively demonstrable. The quantitative study of this precipitation has,
up to the present time, never been done and is very
difficult to achieve. Therefore the question of whether
all IgGs are really precipitable remains open.

Finally, even if absorption is complete, which
fractions are of interest with regards to the pathogenesis of the disease,remains unknown.

Our personal study and those we have mentioned, point
to the fact that in SSPE serum, homogeneous immunoglobulins may be found, that great majority of these immunoglobulins are measles antibodies, and these conclusions are similar to those expressed by Vandvik and
Norrby. We can add a personal observation : these antibodies show a specific idiotypy (17) characteristic for
the patient, rather than for the disease. For the time
being, one can add no more. All these results were
obtained with serum and CSF of SSPE and MS patients. The
studies pertaining to the serum were more successful
for obvious reasons.

From the foregoing, the conclusion can be drawn that
a hyperimmune state exists in SSPE and probably also in
MS. This process leads to the synthesis of homogeneous
measles antibodies in SSPE with specific idiotypy. One
does not know if this reaction is a physiological or a
pathological one.

C. Where does this hyperimmunization reaction occur? Quite
a few observations show that this reaction occurs in the
central nervous system :

1) the anomalies are more distinct in CSF than in serum;
injection of labelled γ globulins made by Frick (18)
showed that these γ globulins reached the CSF in
cases of MS, but that some of the γ globulins are syn-
thesized elsewhere. Research by Cutler (19).

on the vetricular fluid of patients suffering from
SSPE confirms that some of the γ globulins are produced inside the meningeal envelope. The calculation
of ratios for the distribution between serum and CSF
proteins, such as Schuller (20) and Tourtellotte (21)
are using, leads to similar conclusions, demonstrating that the most specific and active immunoglobulins
are relatively more concentrated in CSF than in serum;

2) the study of immunoglobulins produced by CSF cell
cultures (3, 4) confirms this point. These cells
produce immunoglobulins with 'restricted heterogeneity'.
These results were obtained by incorporating radio-
active aminoacids in immunoglobulins produced by CSF
cell cultures and subsequent, electrophoretic examination of these immunoglobulins. One has to remember,
that the patients, who were all MS, had all more than
20 cells per mm3, which is exceptional. We have to
remark that the oligoclonal reaction is seen in CSF in
the absence of cellular reaction, and even in the
absence of any neurological semeiology. Secondly we
think that Sandberg-Wollheim's results should be confirmed by other methods than electrophoresis. Isolated
IgG's easily show as an homogeneous band by electrophoresis.

3) in addition, it was possible to show immunoglobulins
in the nervous tissue by immunohistological methods,
in particular in certain 'glial' cells, which may be
derived from blood.

But the fact that oligoclonal immunoglobulins are
present in the serum, argues against the hypothesis of
of an unique origin in the central nervous system.
Should one admit that a phenomenon, localized in the
central nervous system, could generate enough immunoglobulins to modify the composition of the blood and
even of some organs in SSPE and MS, during years ?
This seems difficult to accept.

In our opinion, we believe that we are confronted

with a general immune response with perhaps a particularly active component in the central nervous system.
This could explain why, so often, these diseases end
clinically with a neurological component. Local
synthesis in the CNS probably occurs. It can be the
consequence of a phenomenon occurring in the nervous
tissue as well as in the cells present in the CSF.
But such a local phenomenon cannot explain the fractionation of serum γ globulins which implicates
necessarily a generalized hyperimmune reaction occu-
ring in the nervous system as well as in other tissues. For us in SSPE, and probably in MS, the immune reaction is a generalized one, better seen or better demonstrated in CSF and located in the CNS as
well as in other tissues.

D. How have we to interprete this hyperimmune reaction?
Is it a continuous and relatively specific stimulation
of antibody synthesis by one or more persistent antigens? This is more than evident, but does not explain
how these antigens can persist in the presence of
their antibodies. Depression of cellular immunity,
the action of a depressive factor which would influence
the cellular reaction and the antigen/antibody reaction,
could be put forward. A depressive T lymphocyte factor
affecting cellular immunity was found by several
authors and was confirmed by the observations made by
Kam-Hansen (1) in the CSF. By precipitating serum IgG
with measles virus total absorption is not achieved (22).
Therefrom the hypothesis was expressed that a factor
present in serum would inhibit this precipitation. This
factor could be a thermostable α 2 globulin and can be
determined by RIA. We have not observed this factor in
CSF. We wonder if we are not in the presence of one and
the same factor which would act simultaneously on the
cellular and the humoral immunity. Maybe it could be
a factor acting on lymphocytes before they differentiate
in T or B lymphocytes.

Has the hyperimmune reaction, as described, to be considered as a physiological, a pathological or ultimately a genetic reaction? It was suggested in the experimental section that hyperimmunization may be influenced by genetic factors. The existence of these genetic factors in SSPE or in MS is still to be proven. Or could it be a pathological reaction, the pathology probably being due to anomalies brought about by the inhibiting factor mentioned before? The fact that free κ and λ light chains (23) were found in CSF of those patients, argues in favour of a pathological immune response.

To us, the immune process seen in the CSF and serum in slow viral diseases, such as SSPE, is a hyperimmune reaction leading to the formation of homogeneous specific antibodies. This reaction can develop thanks to the presence of a factor inhibiting the immune mechanism in serum.

The immune reaction due to the existence of the antigen and the presence of an inhibiting factor is a pathological one. This hyperimmune process may extend to antibodies other than the one mentioned above. Although in SSPE the antigens are to be found in the measles virus, for MS we have no knowledge of the identity of the antigen. Autoimmune reactions against myelin proteins or even IgG (22) were mentioned in MS as well as antibodies against many different viral proteins.

We wish to add the following remarks :

1. The results obtained by the determination of IgM. For IgM (6, 7, 8, 9, 10, 11, 12) and even for IgA (8, 12), a local synthesis is suggested. The variations of the IgM levels might follow closely the clinical phenomena of exacerbations and remissions. The results of these determinations should be accepted with reservation. Determination of IgM by immune complex assays could be an answer to the technical problem. Let us remember

that the changes of IgM may not be correlated with those observed for IgG. The changes in the IgM would be more frequent in SSPE CSF than in MS CSF.

2. The immune complex assays (24). Here as well the re-sults are equivocal. It is not excluded that, when the technical problems have been solved, we might come to interesting conclusions.

3. In spontaneous and experimental spongiform encephalo-pathies, electrophoresis and immune reactions remained normal in CSF and serum.

4. In EAE increase of γ globulins or IgG were reported in CSF. As far as we know, electrophoretic examina-tions were never carried out.

CONCLUSION.

We may conclude by saying that, in at least one of the diseases, considered as slow viral disease, SSPE, the hypo-thesis that the immune reaction observed in CSF is due to a generalized hyperimmunization provoked by a persistent antigen, is accepted. This hyperimmunization leads to the production of homogeneous and relatively specific antibo-dies. It is not excluded that under the influence of a persistent antigen, the humoral system is stimulated in its entirety and that other antibodies are also more actively produced than in normal or basic conditions. We ought to speak here about a generalized relatively specific hyper-immunization. Generalization of the humoral reaction could explain the large number of different IgG fractions observed after electrophoresis and isoelectric focusing. Persistence of the pathological antigen responsible for the hyperimmunization could be explained by the protection of this antigen against the antibodies by a specific inhibitor of the immune reaction. It seems that a genetic factor does not influence the immune reaction. The presence of κ and λ free chains allow to consider the hypothesis that the immune reaction has a pathological character and only is observed in predisposed individuals, for still unknown reasons.

SUMMARY

By electrophoretic study of the CSF proteins we could, for
the first time, demonstrate the restricted heterogeneity
of the IgG. This restricted heterogeneity can be explained
as a hyperimmunization due to the persistance in patients
of an antigen protected against immune reactions by a
factor inhibiting the antigen/antibody reaction. The
immune phenomenon as a whole, would only be observed in
specific individuals. This would explain the rarity of
some of these diseases, such as SSPE.

REFERENCES

1. Kam-Hansen, S: Active T cells in blood and CSF in MS and controls. In: Humoral Immunity in Neurological Diseases, Karcher, D, A Lowenthal, D Strosberg (Eds), Plenum Press, London, 1979, (in press).

2. Lowenthal, A: Agar gel electrophoresis in neurology, Elsevier Publishing Co., Amsterdam, 1964.

3. Sandberg-Wollheim, M: Immunoglobulin synthesis in vitro by cerebrospinal fluid cells in patients with multiple sclerosis. Scand J Immunol 3: 717-730, 1974.

4. Sandberg-Wollheim, M: Immunoglobulin synthesis in vitro by cerebrospinal fluid cells in patients with meningoencephalitis of presumed viral origin, Scand J Immunol 4: 617-622, 1975.

5. Vandvik, B, JB Natvig, D Wiger: IgG 1 subclass restriction of oligoclonal IgG from cerebrospinal fluids and brain extracts in patients with multiple sclerosis and subacute encephalitides, Scand J Immunol 5, 427-436, 1976.

6. Ehrenkranz, NJ, ES Zemel, C Bernstein, K Slater: Immunoglobulin M in the cerebrospinal fluid of patients with arbovirus encephalitis and other infections of the central nervous system. Neurology 24: 976-980, 1974.

7. Schuller, E, N Delasnerie, M Hélary, M Lefèvre: Serum and cerebrospinal fluid IgM in 203 neurological patients. Eur Neurol, 17: 77-82, 1978.

8. Nerenberg ST, R Prasad, ME Rothman: Cerebrospinal fluid, IgG, IgA, IgM, IgD, and IgE levels in central nervous system disorders. Neurology 28: 988-990, 1978.

9. Ziola B, T Vuorimaa, A Salmi, M Pancluis, P Halonen, T Arnadottir, G Enders: Measles IgM antibody persistence: Is it a marker for chronic measles virus infections? Acta Neurol Scand 57, suppl 67: 237-238, 1978.

10. Sindic C, C Cambiaso, PL Masson, EC Laterre: Determination of IgM in the cerebrospinal fluid by particle counting immunoassay (PACIA). In: Humoral Immunity in Neurological diseases, Karcher D, A Lowenthal, D Strosberg (Eds), Plenum Press, London, 1979, (in

press).

11. Olsson JE, H Link: Immunoglobulin abnormalities in multiple sclerosis. Arch Neurol 28: 392-399, 1973.

12. Mingioli ES, W Strober, WW Tourtellotte, JN Whitaker, DE McFarlin: Quantitation of IgG, IgA and IgM in the CSF by radioimmunoassay. Neurology 28: 991-995, 1978.

13. Nordal HJ, B Vandvik, E Norrby: Oligoclonal virus antibodies in healthy adults and neurological patients. In: Humoral Immunity in Neurological Diseases, Karcher D, A Lowenthal, D Strosberg (Eds), Plenum Press, London, 1979, (in press).

14. Lowenthal A, K Adriaenssens, B Colfs, D Karcher, R Van Heule: Oligoclonal gammapathy in ataxia telangiectasia. Z Neurol, 202 : 58-63, 1972.

15. Strosberg AD, D Karcher, A Lowenthal: Structural homogeneity of human subacute sclerosing panencephalitis antibodies. J Immunol, 115, 157-160, 1975.

16. Karcher D, G Matthyssens, A Lowenthal: Isolation and characterization of IgG globulins in subacute sclerosing panencephalitis. Immunology, 23: 93-99, 1972.

17. Strosberg AD, D Karcher, A Lowenthal: Structure and idiotype of human homogeneous antibodies. Immunology 4249, Fed. Proc 969, 1975.

18. Frick E, L Scheid-Seydel: Untersuchungen mit I^{131} markiertem γ -Globulin zur Frage der Abstammung der Liquoreiweisskörper. Klin Wschr 36: 857-863, 1958.

19. Cutler RWP, GV Walters, CF Barlow: ^{125}I-labelled protein in experimental brain oedema. Arch Neurol 11, 225-238, 1964.

20. Schuller E: Cerebrospinal fluid immunoglobulins. II. Variations in different pathological situations. La nouvelle presse médicale, 8: 427-432, 1979.

21. Tourtellotte WW: Cerebrospinal fluid in Multiple Sclerosis. In: Handbook of Clinical Neurology, Vinken PJ, GW Bruyn (Eds), North Holland Pub, Amsterdam, 1970, p 324-382.

22. Karcher D, M Noppe, A Lowenthal: A heat stable serum inhibitor of an antigen antibody reaction of subacute sclerosing panencephalitis. J Neurol, 261/1: 51-56, 1977.

23. Bollengier F, A Lowenthal, N Henrotin: Bound and free light chains in SSPE and MS serum and CSF. Z klin Chem, 13: 305-310, 1975.

24. Masson PL: Circulating immune complexes in neurological
    disorders. In: Humoral Immunity in Neurological Di-
    seases. Karcher D, A Lowenthal, D Strosberg (Eds),
    Plenum Press, London, 1979, (in press).

DISCUSSIONS OF PAPERS BY A. LOWENTHAL AND D. KARCHER

Dr Kreth commenced discussion of this paper by pointing out that hyper-immunisation in SSPE was probably a pathological process and not a normal one. There was little evidence for a possible failure of measles antigen in the peripheral circulation to form immune complexes. It was therefore not known how measles antibody production in SSPE took place _in vivo_;   certainly, there was some evidence for high antibody levels before the onset of clinical signs. It might be that there was something wrong with the regulation of antibody production and that SSPE patients could not stop or regulate the production of antibody and this production had to go on and on. Such a regulatory error might involve interactions between regulatory and suppressor T cells and might be extremely specific. It could be related to the degree to which the patient had been exposed to the antigen. It was noted that, in SSPE patients, plasma exchange had no effect and the antibody levels rose extremely rapidly in spite of this procedure.

Dr Norrby asked why one should invoke a congenital immunological abnormality when there were clearly large amounts of antigen present. The position differed from MS, where the immune reaction was important. In SSPE, 80% - 90% of the immunoglobulin was directed against measles antigen but, in MS, the proportion was very much lower, most of the immunoglobulin being directed against other antigens.

INFECTIONS IN IMMUNODEFICIENT PATIENTS

A. D. B. WEBSTER

Introduction

Antibodies protect humans against many virus infections. This is shown
by the finding that patients with primary hypogammaglobulinaemia who
are given regular gammaglobulin injections, do not suffer from the
common virus infections of childhood such as measles, varicella and
mumps. However, the mechanisms involved in eliminating established
virus infections are not well understood. The traditional view is
that cell mediated immunity plays a predominant role, probably though
the killing of virus infected cells (cytotoxicity) by lymphocytes and
macrophages. Nevertheless, the fact that some patients with primary
hypogammaglobulinaemia develop chronic echovirus infection suggests
that, in this instance, antibodies may also be important.

The virus infections which occur as complications of hypogammaglobulinaemia should not be confused with the virus infections which
cause immunodeficiency. Both foetal rubella (1) and infectious mononucleosis in children(2) may cause hypogammaglobulinaemia but the infection does not persist in an overt form.

This chapter will describe the virus infections which occur in
patients with primary defects of cellular or humoral immunity. Virus
infections are not a recognised complication of the rare primary defects
in neutrophils or the complement system.

PRIMARY ANTIBODY DEFICIENCY

SEX LINKED HYPOGAMMAGLOBULINAEMIA

This rare inherited disease is characterised by recurrent bacterial
infections of the upper and lower respiratory tract, meningitis and
occasionally septic arthritis in the first two years of life (3). The
organisms involved are usually pneumococci and H.influenzae although
they are also prone to mycoplasma infections. Virus infections,
except for those due to certain enteroviruses, are not a troublesome
complication.

The patients have very low levels of serum immunoglobulins and are unable to make antibody when immunized. A characteristic feature, from the laboratory point of view, is the complete absence of circulating B lymphocytes. However, they do have cells in the bone marrow which are thought to be the precursors of B cells (4). Cellular immunity is intact, the patients showing normal delayed hypersensitivity skin reactions, macrophage inhibition factor production and normal in vitro lymphocyte transformation to mitogens and allogeneic cells. Their circulating T lymphocytes have normal 'natural' non-specific cytotoxicity against various target cells (5). Cytotoxicity against antibody coated cells, such as chicken erythrocytes, is also normal ('K' cell killing) but the killing of antibody coated allogeneic lymphocytes is severely depressed (6). This finding suggests that the latter type of cytotoxicity is mediated by a B lymphocyte.

Echovirus infections

In 1956, Janeway (7) described a condition like dermatomyositis in a child with sex-linked hypogammaglobulinaemia. The main features were the gradual onset of brawny oedema of the extremities with flexion contractures of the knee and elbow joints which produced a characteristic posture. In 1971, we investigated an 11 year old boy (AG) with sex-linked hypogammaglobulinaemia who presented with a two months history of severe headaches. He then developed brawny oedema of the arms with some swelling of the elbow and wrist joints (8). Echovirus 11 was cultured from his cerebral spinal fluid which also contained raised protein and mononuclear cells. A skin and muscle biopsy showed marked perivascular cuffing by mononuclear cells. There was relatively little muscle fibre destruction which explained why the serum creatinine phosphokinase level was normal. Although the oedema in his arms rapidly improved after treatment with human immune plasma, his headaches persisted and he had frequent grand mal convulsions. Central nervous system features were variable and, during good phases, he was able to return to school. He developed hydrocephalus six months after the start of his illness which required an atrial ventricular shunt. Two years later, he died suddenly with respiratory centre failure.

The autopsy on patient AG showed a severe chronic meningitis

with thickening of the leptomeninges of the brain and spinal cord.
There were scattered destructive lesions in the brain showing marked
astrocytic proliferation, microcalcification and perivascular lymphocytic infiltration.   There was depletion of neurones in some parts
of the central grey matter, cerebella nuclei and cerebellum.

Wilfert et al (9) reported six similar cases in the U.S.A.
Echovirus was isolated from the cerebral spinal fluid of all patients
although not all had central nervous system features.   A subclinical
hepatitis was common.   Headaches, deafness and convulsions occurred
in some patients and one had iridocyclitis.   Although only two of the
patients in this series had died at the time of reporting, it is
generally believed that the disease is nearly always ultimately fatal.

Table 1 shows the features of the three patients we have managed.
Patient WF developed episodes of erythema and swelling of the lower
legs two years before he complained of headaches and sensory disturbances.   His central nervous system disease progressed despite treatment with immune serum and he died of respiratory centre failure
shortly after a bone marrow graft.   MR was noticed to have flexion
deformities of his elbows and knees at a routine clinical examination.
Echovirus 17 was cultured from his cerebral spinal fluid, which also
had a high protein and cell count, despite the absence of central
nervous features.

Age at present- ation of echo- virus infection	Echovirus type isolated from CSF	Clinical features	Outcome
24	3	Oedema of lower legs, deafness, headaches, sensory disturbances	Died after 3 year ill- ness of respiratory centre failure. Attempted bone marrow graft.
11	11	Headaches, deafness oedema of arms grand mal convulsions hydrocephalus	Died after $2\frac{1}{2}$ year illness of respiratory centre failure
18	17	Flexion deformities of elbows and knees, deafness	Condition unchanged after 18 months

Footnote: All three patients had severe congenital hypogammaglubulinaemia
(two with affected male relatives) with absent circulating B lymphocytes
and normal cellular immunity).

Treatment

All three of our patients have been treated with hyperimmune animal
serum, patient AG receiving repeated infusions of horse serum while
patients MR and WF were given sheep serum. The horse serum was
obtained commercially but we raised our own immune sheep serum. This
was done by propagating the echovirus from each patient on MRC 5
fibroblasts in an isolated laboratory. The tissue culture fluid was
then processed to produce an inactivated vaccine by methods similar
to those used for the production of polio vaccine. The sheep were
immunized with the vaccine in Freunds complete adjuvant at multiple
intradermal sites. Booster injections of vaccine in Freunds incomplete
adjuvant were given at monthly intervals until an adequate antibody
titre was reached. Most sheep produced a serum antibody titre of
> 1:512 (measured by haemagglutination inhibition or neutralization)
within about two months.

Two of the patients received about 200 ml of whole immune sheep
serum intravenously every 2 - 3 weeks. Despite treatment for one
year, the central nervous system features in patient WF progressed
until his death. The swelling of his lower legs also persisted. MR
received similar therapy for about a year and a repeat muscle biopsy,
taken three months after starting serum therapy, was much improved.
The flexion deformities of his elbows and knees have persisted. For
the last seven months, we have been unable to isolate echovirus from
his cerebral spinal fluid although the last two specimens have
contained a virus-like agent (see Tyrrell's chapter). The dermatomyositic features in patient AG improved rapidly when he was given
human plasma containing anti-echovirus 11 antibody at a titre of 1:80
(haemagglutination inhibition test). However, his central nervous
system disease progressed despite treatment with hyperimmune horse
serum, maternal peripheral blood white cells taken after she had been
immunized with an echovirus 11 vaccine, and various attempts to non-
specifically stimulate cellular immunity (i.e. BCG vaccination,
transfer factor, Lamprene). Patient AG initially showed evidence of
non-specific depression of cellular immunity with poor lymphocyte
transformation and absence of delayed hypersensitivity skin reactions.
However, these returned to normal with improvement in his general
condition and we were able to demonstrate specific immunity against

echovirus 11 using a macrophage inhibition test (8).

Our experience indicates that the dermatomyositic features of this disease can be improved by treatment with hyperimmune animal serum. However, there is general agreement in both the U.S.A. and this country that such therapy does not alter the progression of the central nervous system disease. Since very little antibody is likely to cross the blood brain barrier, a more rational approach would be to give intra-thecal hyperimmune globulin. Such therapy has been used successfully in the treatment of neonatal tetanus (10). It is probably dangerous to inject whole animal serum intrathecally as this commonly produces local inflammation when given intravenously. However, purified hyperimmune sheep globulin would probably be safe. Interferon therapy should also be considered but this would have to be given intrathecally as it does not cross the blood brain barrier.

Bone marrow transplantation is the only method currently available for reversing the immunological defect. However, in the absence of a histocompatible sibling, this is not worth consideration. The immunosuppression given to patient WF during his bone marrow transplant from his histocompatible sister did not seem to cause dissemination of the echovirus infection, and his death from respiratory centre failure a few weeks later was probably fortuitous.

Prophylaxis

The gammaglobulin therapy given to these patients usually offers no protection against echovirus infection. This is because the gamma-globulin is prepared one or two years before it is given to the patient and therefore does not contain antibody to the prevalent echovirus serotypes in the community at the time. An alternative approach would be to produce hyperimmune sheep globulin against a wide range of common echovirus serotypes. This could then be given at roughly monthly intervals to susceptible patients. The finding that three out of about 15 patients with either proven or probable sex-linked hypogammaglobulinaemia have developed echovirus infection, shows that this is a common complication and that it would be economically sensible to invest in such a prophylactic regime. Such therapy would only need to be given to those patients with childhood onset hypogammaglobulinaemia with absent circulating B lymphocytes.

Poliovirus

The incidence of paralytic poliomyelitis due to natural infection is
probably not raised in patients with primary hypogammaglobulinaemia.
However, there does seem to be a raised incidence of vaccine
associated poliomyelitis and Wright et al (11) reviewed five cases.
The disease usually has a prolonged incubation period of more than
two months and at least two cases have survived with only minor
muscle weakness. This complication does not occur in patients
already receiving gammaglobulin therapy and none of the 28 children
in the Medical Research Council series (12), who were given oral polio
vaccine, developed paralytic disease. However, two of these patients
continued to excrete the vaccine strain in the stools for up to 32
months.

ADULT ONSET 'VARIABLE' HYPOGAMMAGLOBULINAEMIA

Herpes zoster infection occurred in 18% of the patients in our series.
This nearly always affected patients who were not receiving gammaglobulin therapy and was self-limiting and localised. Some patients
have had two or three attacks involving different dermatomes. This
high incidence of Herpes zoster may be explained by the finding that
many of these patients have defects in cellular immunity as shown by
absent delayed hypersensitivity skin reactions and poor in vitro
lymphocyte transformation (13).   The common occurrence of Herpes
zoster in Hodgkin's disease, and other lymphomas associated with
similar defects in cell mediated immunity, supports this view.
However, not all the patients with Herpes zoster in our series have
demonstrable defects in cellular immunity. Herpes zoster is not a
complication of sex-linked hypogammaglobulinaemia so that the antibody
deficiency alone cannot be responsible.

Herpes simplex (labialis) is extremely rare in patients with
hypogammaglobulinaemia regardless of whether they are receiving
gammaglobulin therapy.

SELECTIVE DEFECTS IN CELLULAR IMMUNITY

There is a very rare group of patients with T lymphocyte deficiency.
This may be caused by a non-familial foetal abnormality causing

absence of the thymus (14) or to an inherited deficiency of the enzyme,
nucleoside phosphorylase (15).   There is very little information
concerning virus infections in thymic aplasia but patients with
nucleoside phosphorylase deficiency are prone to severe, and sometimes
fatal, varicella and cytomegalovirus infections.   For these reasons,
prophylactic gammaglobulin injections should be given to these
patients.

SEVERE COMBINED IMMUNODEFICIENCY

This disease is characterised by a susceptibility to bacterial,
fungal and viral infections, usually starting within the first few
months of life   (16).   Affected infants fail to thrive and usually
die within the first two years unless given a successful bone marrow
graft.   There is an autosomal and sex-linked recessive form of the
disease, one cause for the latter variety being a deficiency of the
enzyme, adenosine deaminase (17).

Affected children are prone to a variety of viruses such as
adenovirus, cytomegalovirus and measles.   They frequently suffer
from unexplained diarrhoea which may be caused by viruses although
they are rarely isolated.   Viral encephalitis is common and viruses
such as echovirus, Herpes simplex and measles virus have been
isolated from the brains of some children at autopsy.   Dayan (18)
found histological evidence of viral encephalitis in 9 of 23 brains
examined at autopsy from children with severe combined immunodeficiency.   Immunization with live viral vaccines is hazardous and
generalised fatal vaccinia is a well-known complication.   There is
also a raised incidence of vaccine-associated paralytic poliomyelitis which is usually fatal (19, 20).   Nevertheless, a few
patients have received live oral polio vaccine without complication
(21).

CONCLUSION

The finding that children with severe defects in T lymphocyte
function often die from viruses supports the view that cellular
immunity is an important defence mechanism. However, this does not
seem to apply to echoviruses where either antibody or B lymphocytes
are required to eliminate the infection. The chronicity of the
central nervous system infection in patients with hypogamma-

globulinaemia, even in the absence of neurological signs, should
encourage a search for viruses in other chronic inflammatory
diseases of the central nervous system.   These patients also
provide a challenge to find a way of treating chronic virus infections
of the central nervous system.

REFERENCES

1. Soothill, JF, K Hayes, JA Dudgeon: The immunoglobulins in congenital rubella. Lancet 1: 1385-1388, 1966

2. Purtilo DT, JPS Yang, S Allegra, D De Florio, LM Hutt, M Soltani, G Vawter: Hematopathology and pathogenesis of the X-linked recessive lymphoproliferative syndrome. Am J Med. 62: 225-233, 1977

3. Webster ADB: Immunodeficiency. In: Medical Immunology, Holborow EJ and WG Reeves (eds), Academic Press, 1977, p474-537

4. Pearl ER, LB Vogler, AJ Okos, WM Crist, AR Lawton, MD Cooper: B lymphocyte precursors in human bone marrow: an analysis of normal individuals and patients with antibody-deficiency states. J.Immunol 120: 1169-1175, 1978.

5. Koren HS, DB Amos, RH Buckley: Natural killing in immuno-deficient patients. J.Immunol. 120: 796-799, 1978

6. Sanal SO, RH Buckley: Antibody-dependent cellular cytotoxicity in primary immunodeficiency diseases and with normal leukocyte subpopulations: importance of the type of target. J.Clin.Invest. 61: 1-10, 1978

7. Janeway CA, D Gitlin, JM Craig, DS Grice: 'Collagen disease' in patients with congenital agammaglobulinaemia. Trans. Ass. Am. Physicians 69: 93-97, 1956

8. Webster ADB, JH Tripp, AR Hayward, AD Dayan, R Doshi, EH MacIntyre, DAJ Tyrrell: Echovirus encephalitis and myositis in primary immunoglobulin deficiency. Arch.Dis.Child 53: 33-37 1978

9.   Wilfert CM, RH Buckley, T Mohanakumar, JF Griffith, SL Katz,
     JK Whisnant, PA Eggleston, M Moore, E Treadwell, MN Oxman
     FS Rosen: Persistent and fatal central nervous system
     echovirus infections in patients with agammaglobulinaemia.
     N.Engl.J.Med.296: 1485-1489, 1977

10.   Sanders RKM, R Joseph, B Martyn, ML Peacock: Intrathecal
      antitetanus serum (horse) in the treatment of tetanus.
      Lancet 1: 974-977, 1977

11.   Wright PF, MH Hatch, AG Kasselberg, SP Lowry, WB Wadlington,
      DT Karzon: Vaccine associated poliomyelitis in sex linked
      agammaglobulinaemia. J.Pediatr.91: 408-412, 1977

12.   MacCallum FO: Antibodies in hypogammaglobulinaemia. In:
      Hypogammaglobulinaemia in the United Kingdom.
      Medical Research Council SRS 310. Her Majesty's Stationery
      Office, London. p.72-85, 1971

13.   Webster ADB, GL Asherson: Identification and function of
      T cells in the peripheral blood of patients with hypogamma-
      globulinaemia. Clin.Exp.Immunol.18: 499-504, 1974

14.   Lischner HW, DS Huff: T-cell deficiency in Di George syndrome.
      In: Immunodeficiency in Man and Animals. Birth Defects:
      original article series, Vol 11, No. 1. p16-21, 1975

15.   Biggar WD, ER Giblett, RL Ozere, BD Grover: A new form of
      nucleoside phosphorylase deficiency in two brothers
      with defective T cell function. J.Pediatr. 92: 354-357, 1978

16.   Rosen FS: Immunity deficiency in children. Br.J.Clin.Pract.
      26: 318-322, 1972

17.   Meuwissen HJ, B Pollara, RJ Pickering: Combined immuno-
      deficiency associated with adenosine deaminase deficiency.
      J.Pediatr. 86: 169-181,1975

18. Dayan A.D: Chronic encephalitis in children with severe
     immunodeficiency.   Acta Neuropathol. 19: 234-241, 1971

19. Davis LE, D Bodian, D Price, J Butler, JH Vickers: Chronic
     progressive poliomyelitis secondary to vaccination of an
     immunodeficient child. N.Engl.J.Med.297:241-245,1977

20. Feigin RD, MA Guggenheim, SD Johnsen: Vaccine related
     paralytic poliomyelitis in an immunodeficient child.
     J Pediatr 79: 642-647, 1971

21. Lopez C, WD Biggar, BH Park, RA Good: Nonparalytic
     poliovirus infections in patients with severe combined
     immunodeficiency. J Pediatr 84: 497-502, 1974.

DISCUSSIONS OF PAPER BY A. D. B. WEBSTER

Dr ter Meulen opened the discussion, asking why there had been no
recoveries from these patients of enteroviruses other than ECHO viruses.
In answer, Dr Webster stated that this was not known; certainly,
Coxsackie virus infections did occur in these patients and there was
a higher incidence of paralysis with vaccine strains of poliovirus.
In connection with the myositis observed in these patients, Dr Tyrrell
commented on the similarity of some strains of ECHO virus (e.g. 9 and
24) to Coxsackie viruses. Following a query concerning the effects
of natural measles virus infection in these patients, it was noted
that, in this country at least, they were usually recognised in early
infancy and treated with prophylactic gamma-globulin, which contained
measles virus antibodies.

Dr Cathala stressed that the disease in these patients was a
chronic meningitis and not a neurological disease. In a final
exchange, a postulated comparison between the effects of dengue virus
infection and the conditions which Dr Webster had described was not
upheld.

VIRUS-LIKE AGENTS FROM PATIENTS WITH MENTAL DISEASES AND SOME CHRONIC NEUROLOGICAL CONDITIONS

D. A. J. TYRRELL, R. PARRY, T. J. CROW, E. JOHNSTONE AND N. FERRIER

My colleagues and I have described how a virus-like cytopathic effect (CPE) was detected in cultures inoculated with CSF collected from patients with schizophrenia and some chronic progressive nervous disease, such as Huntington's chorea (1, 2). The main observations were that within a few days of inoculation a focal CPE was seen, particularly in stationary unchanged cultures of the MRC-5 strain of human embryo lung fibroblasts incubated at 33°C. The CPE tended to disappear and could only rarely be passed serially. However the agent causing it was partly characterised by experiments on CSF and it was shown to be particulate by filtration, and also appeared to resist lipid solvents, heat at 56°C and grew in cultures treated with BUDR. It (or they) has been therefore called provisionally a virus-like agent(s) or VLA. Such observations prompt us to ask a series of questions, to which at the moment we have only a few answers, but this paper gathers together our more recent observations and our ideas under the headings of these questions.

IN WHAT CLINICAL CONDITIONS CAN WE DETECT VLA?
We have received carefully collected and stored CSF from a variety of patients. The most nearly "normal" fluids came from patients undergoing spinal anesthesia for conditions such as benign prostatic hyperplasia, and from patients who had myelograms or air studies for conditions such as backache in which no abnormality was found. We also had some patients with general medical conditions such as cardiac arrest, or septicaemia, and with acute nervous system infections such as bacterial aseptic meningitis or unexplained encephalitis. The present catalogue of results is shown in table 1, which indicates that such specimens are usually negative but can be positive in patients with acute neurological syndromes. One patient with hypogammaglobulinaemia had suffered from an echovirus infection which had subsided following intensive specific immunotherapy (3).

Table 1   Results of tests on CSF from certain
surgical and medical cases

Clinical features of cases	Proportion tested showing CPE
Lumbar puncture for spinal anaesthesia or for radiological study of backache[*]	3/22
Various mainly infectious intracranial diseases - stroke, convulsions, subarachnoid haemorrhage or cranial nerve palsy	1/12
Meningism or meningitis	1/7
Encephalitis	0/3
Cardiac arrest, septicaemia, headache, vomiting	0/6
Acute unexplained fluctuating loss of consciousness	3/3
"Hysterical" hemiplegia	1/1
X-linked hypogammaglobulinaemia with CNS involvement	1/1
"Sea-blue" histiocytosis	1/1

* The clinical records on some of these patients are incomplete
  - some patients did have neurological syndromes.

Table 2   Tests on cases of psychiatric illness,
            Huntington's chorea and multiple sclerosis

	Proportion tested showing CPE
Schizophrenia	10/14
Affective disorders	
Hypomania	2/3
Depression	3/3
Confusional state	1/1
Severe anxiety state	*0/1
Huntington's chorea	3/4
Multiple sclerosis	4/8

* A VLA was detected by throat swab but the CSF was negative.

However, we have evidence that VLA may be detected in multiple
sclerosis, particularly in those with active or progressive disease
(table 2) and also in Huntington's chorea. We have been interested
to detect VLA in several cases with what might be labelled acute
confusional or organic mental states; some patients had mild respiratory or influenza-like infections and signs suggestive of CNS
involvement such as meningism and headache - three were drowsy or
apparently unconscious. Our original study concentrated on patients
with schizophrenia and we now detect VLA in about two-thirds of cases.
More recently we have begun to study patients with other forms of
mental disease and we have detected VLA in the first small group of
cases of affective disorders. Examples of the CPE are shown in
fig. 1. The clinical features of the schizophrenic patients have
been summarized - those of the other cases will be reported later.

CAN VLA BE CONTINUOUSLY PROPAGATED IN VITRO?

There is some evidence that the CPE is due to a replicating agent.
Not only have a few foci spread gradually through a cell sheet, but
occasionally we have been able to produce CPE in further cultures by
passing the culture fluids to further cultures of MRC-5 or of brain
tumour cells up to a total of three passages. It is clear that the
CPE is usually reduced by using cultures which have been recently
changed or are rolled. In recent months we have had further interesting results. We received tissue from the frontal and temporal cortex
of a recently deceased patient with Huntington's chorea. This was
inoculated into roller tissue cultures. A cytopathic effect developed
in rhesus monkey kidney cells and can now be propagated serially in
HeLa cells. It is now necessary to study the agent in detail to
determine whether it is related to the disease. We cannot exclude
that it might have been derived from the rhesus kidney cells. A
similar CPE was seen in cultures inoculated with throat swab material
from a patient with schizophrenia and this was at first thought to be
due to an adenovirus. This agent likewise will be characterized and
using similar techniques we shall continue our efforts to get a
continuously propagated agent from at least representative cases of
other types of disease. It thus seems that by persistence an
occasional VLA can be persuaded to grow serially, but great care

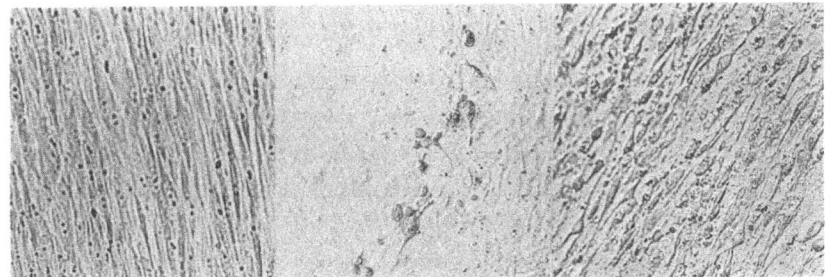

Fig. 1 Cytopathic effect produced in human embryo lung fibroblast
cultures (MRC-5) inoculated with CSF. Unstained.
a) Uninoculated culture
b) Culture inoculated 2 days previously with 0.1 ml of CSF
from a patient with Huntington's chorea. An early focus.
c) Culture inoculated 4 days previously with 0.1 ml of CSF
from a patient with multiple sclerosis. An extensive
cytopathic effect.

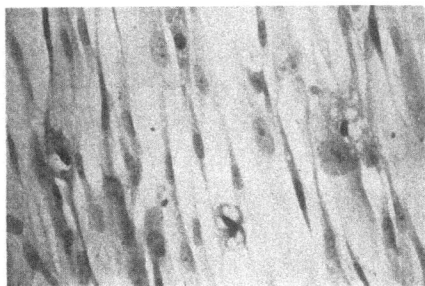

Fig. 2 Fixed and stained MRC-5 cells inoculated with CSF from
patient with schizophrenia. Note the vacuolation of the
cytoplasm and nuclear changes. Stained haematoxylin and
eosin.

will have to be exercised to ensure that what is grown really is a
relevant organism and not a passenger virus or laboratory contaminant.

WHAT IS THE VIROLOGICAL SIGNIFICANCE OF THESE OBSERVATIONS?

We have considered the possibility that these results were spurious
and due to the use of unsatisfactory cells or to faulty observations.
We think these are unlikely explanations for we have observed many
hundreds of uninoculated cultures and only rarely seen any changes
comparable with the CPE, and in many instances the CPE is so marked,
rapid and extensive that untrained observers have readily recognized
it. Furthermore many of the observations were made by observers who
did not know the origin of the specimen and in each test there have
been specimens from a mixture of cases. We have often passed uninoculated cells and have not seen a comparable CPE, suggesting that
it is unlikely to be an agent latent in our cultures. Furthermore
the strongly positive specimens have been positive in HeLa and monkey
kidney cells as well as in MRC-5 cells. Some of our cultures contain
mycoplasmas and we also get occasional contamination with fungi,
bacteria or protozoa, but the properties of the agent deduced by
tests on CSF are apparently incompatible with a contaminating mycoplasma or other organism. After some initial studies we arranged
for all the laboratory manipulations of our VLA to be confined to
one exhaust ventilated cabinet into which known viruses were never
introduced - this was intended to minimize the risk that virus was
carried across from other cultures. The fact that the agent cannot
be serially passed makes it unlikely that it is due to a laboratory-
grown virus, and contamination from positive specimens is unlikely
because the titre of these is low and numerous uninoculated control
cultures have never shown the effect. Thus the best hypothesis is
still that we are seeing the effect of a virus-like agent in the
patient's CSF.

We would like to know whether we are dealing with one agent or
several. The properties, so far as they go, suggest no obvious
differences though one agent from a case of "sea blue" histiocytosis
with choreo-athetosis is apparently chloroform labile. The CPE
varies in scale and also in the speed with which it develops, but
qualitatively it seems to be the same whatever the type of case and

in unstained cultures resembles that due to viruses, for example
rhinoviruses. No characteristic inclusions have been seen and the
changes seem to be mainly vacuolation of the cytoplasm and coarsening
of the nuclear chromatin, followed by pyknosis or karyorrhexis
(fig. 2). Eventually one hopes this will be supported by more
detailed and specific studies such as immunofluorescence and electron
microscopy. Nevertheless we could be dealing with several closely
related viruses, for instances a number of serotypes of a single
agent, or even quite unrelated agents - we should remember that at
an equally early stage of research the reoviruses were grouped with
the enteroviruses and only later was echovirus type 10 renamed reo-
virus type 1.

WHAT IS THE RELATIONSHIP OF VLA TO DISEASE?
We wondered at first whether VLA was abnormal but previously undet-
ected inhabitant of the CSF. Failure to recognize a normal symbiont
as such has mislead investigators in the past. It is not ethical to
collect CSF from completely normal subjects, but we are impressed
that generally we have not found VLA in CSF of patients with diseases
unrelated to the CNS, such as mild orthopaedic disabilities, surgical
conditions such as benign prostatic hypertrophy and even serious
general medical conditions such as septicaemia and cardiac arrest.
We could postulate, of course, that VLA is released by damaged nervous
tissue, and might be a non-specific indicator of damaged nervous
tissue. However, samples of CSF from patients with severe and
extensive acute CNS disease, such as meningitis and encephalitis,
were uniformly negative.

    Thus probably the simplest hypothesis is that VLA is aetiologic-
ally associated with some CNS diseases including schizophrenia,
affective psychoses, Huntington's chorea and multiple sclerosis.
It is regularly found in close association with the diseased tissue
and not in unaffected CNS, and it is capable of damaging cultured
nervous system cells as well as a variety of others. However the
evidence would be much reinforced if we could produce CNS disease
by inoculating VLA into the CNS of animals, and experiments of this
type have now begun. If they are positive they will fulfil the
remaining third clause of Koch's postulates.

We have some evidence that VLA may cause mild and recoverable
conditions. For instance there is the small group of adult patients
who had rather rapid impairment or loss of consciousness without any
explanation apart from signs suggestive of a mild infection, and two
of these recovered rapidly and completely. We have also found VLA
in the CSF of an elderly man with backache and of a 14 year old girl
who appeared to have a respiratory infection with meningism, followed
by an attack of sinusitis - again both of these recovered.

IF VLA IS PROVED TO BE IMPORTANT WOULD THIS VITIATE RESULTS WHICH
SUGGEST THE IMPORTANCE OF GENETIC AND ENVIRONMENTAL FACTORS IN THESE
DISEASES?

In most of the diseases in which we have found VLA there is little
or no evidence of case association, let alone case-to-case transmission of an infectious agent. The best hypothesis would appear
to be that VLA is widespread in the community. Since we have positive
results with two throat swabs and one faeces it is not difficult to
imagine how it could be transmitted without involving the CNS at all.
Genetic predisposition might then determine whether clinical illness
occurs, as in the case of Huntington's chorea. Nevertheless it is
completely obscure at the moment how a genetic influence might
operate - it could affect the susceptibility of cells to the virus
or the ability of the immune system to respond.

On the other hand in schizophrenia probably other factors
besides the genetic predisposition would be important - for example
the age at which infection occurs could have an important influence.
Then the content of a schizophrenic illness would be very much
influenced by the social and educational background; for instance
these must determine which language hallucinating voices speak, and
whether the influences are conceived of as coming from outer space.
It must also be clear that the changes in the brain can be described
in many ways, in terms of electrical activity, or the types of
neurones or receptors affected, e.g. dopaminergic. Nevertheless
the changes so described could all be the consequence of a virus
invading a series of susceptible cells.

Others have considered the possibility that schizophrenia might
be due to a virus infection and found the idea plausible (4, 5)

though at the time in most cases there was no direct evidence to support the idea.

The pathogenesis could still be complicated and be partly mediated by immune processes. It could perhaps have something in common with the long term damage produced in the mouse CNS by Theiler's virus, a picornavirus (6, 7) or in the so called immune polioencephalitis which occurs in old mice, particularly of the C58 strain, and now known to be transmitted by a probable togavirus (8).

CONCLUSION

This can be only a preliminary report, a sketch map of what we think may well be an exciting new land, an area of knowledge in which we can explore the interactions between some new organisms and the central nervous system of man. Clearly much more work is needed so that details may be filled in and many unknown facts discovered.

ACKNOWLEDGEMENTS

We wish to thank again the many individuals referred to in our recent publication, without whose efforts this research would not have been possible.

REFERENCES

1. Tyrrell, DAJ, RP Parry, TJ Crow, E Johnstone, IN Ferrier:
   Possible virus in schizophrenia and some neurological
   disorders. Lancet i: 839-841, 1979.
2. Crow, TJ, IN Ferrier, EC Johnstone, JF Macmillan, DGC Owens,
   RP Parry, DAJ Tyrrell: Characteristics of patients with
   schizophrenia or neurological disorder and virus-like agent
   in cerebrospinal fluid. Lancet i: 842-844, 1979.
3. Webster, ADB, JH Tripp, AR Hayward, AD Dayan, R Doshi,
   EH Macintyre, DAJ Tyrrell: Echovirus encephalitis and
   myositis in primary immunoglobulin deficiency. Arch. Dis.
   Childh. 53: 33-37, 1978.
4. Torrey, EF, MR Peterson: Slow and latent viruses in schizophrenia. Lancet ii: 22-24, 1973.
5. Crow, TJ: Viral causes of psychiatric disease. Postgrad. med.
   J. 54: 763-767, 1978.
6. Theiler M: Spontaneous encephalomyelitis of mice, a new virus
   disease. J. exp. Med. 65: 705-719, 1937.
7. Lipton HL, MC Dal Canto: Contrasting effects of immunosuppression in Theiler's virus infection of mice. Infect. Immun.
   15: 903-909, 1977.
8. Martinez D, B Wolanski, AA Tytell, RG Devlin: Viral aetiology
   of age-dependent polioencephalomyelitis in C5B mice. Infect.
   Immun. 23: 133-139, 1979.

DISCUSSION OF PAPER BY D. A. J. TYRRELL ET AL.

Discussion on Dr Tyrrell's paper opened with a question on the use of immunosuppressed animals in future investigations. Dr Tyrrell replied that experiments in a wide range of animals were contemplated and reminded the audience of a disease of the mouse central nervous system which was initially thought to be immunological in origin. Here, only old mice, or mice which had been treated with cyclophosphamide, were susceptible. In due course, it appeared that a toga-virus was responsible.

Dr Crow mentioned the use of behavioural alterations (already being used as a model system in the investigation of schizophrenia) in the detection of mild degrees of illness. He suggested the testing of brains from such animals in tissue culture. As these virus-like agents are now capable of recovery from faeces and throat swabs, it now seemed feasible to test for their presence in unaffected normal subjects.

Dr Thiry suggested that these virus-like agents might resemble the adeno-associated viruses. Dr Tyrrell, in reply, drew attention to the limited host range of the virus-like agents. Of 7 clones of cells from one foetus, only one was suitable for the culture of these agents. It was also noted that transformed MRC-5 cells were resistant to infection with these agents. Professor Chany recalled that, in 1960 - 1961, Dr Gresser, working in his laboratory, identified similar agents which could not be successfully passaged.

In answer to further questions, Dr Tyrrell stated that the cytopathic effect developed usually within 2 or 3 days (often 24 hours) and that the stage in the disease at which specimens had been obtained varied from 3 days to more than 20 years. In further discussion, it appeared that there was a significant literature on the development of schizophrenia following virus infections and that, in 1918 - 1919, schizophrenia followed the influenza pandemic. Vilyuisk encephalitis could present as a schizophrenic illness which progressed to a dementia.

In a discussion of technical details, it was apparent that the cytopathic effect only appeared if the conditions were just right; thus immunofluorescence with convalescent sera and similar procedures were not possible as satisfactory cover-slip preparations could not be made. The agents were sensitive to freezing (-40°C) and to freezing and thawing but resisted overnight storage at 4°C and could be safely kept for longer periods in liquid nitrogen or at -70°C. These virus-like agents were deposited by ultracentrifugation.

Dr Kimberlin, predicating that this might be a scrapie-like agent, suggested that it should not be inoculated into neonatal animals nor should immunosuppression be used. Professor Fraser observed that these experiments reminded him of ones currently under way in Belfast where a slightly different cytopathic effect had been produced in VERO cells inoculated with human marrow specimens.

In conclusion, it was suggested that a study of close relatives of patients with Huntington's chorea should be set up; it was agreed that such a study would be quite feasible.

DISCUSSION ON DR. SANGER'S PAPER*

This interesting paper led to a lengthy discussion, beginning with a consideration of the mode of spread of infection of viroids. As a point of general interest, the necessity for asymptomatic infection to occur was considered to be important in the understanding of persistent infections. In the case of viroids, the mode of spread from one plant to another was imperfectly understood and the vectors were not known. On the analogy of the experimental means of transfer and the natural spread associated with the use of pruning shears, both of which procedures involve damage to the plant, the possibility of transmission by biting insects was seriously considered but there was no data at all on this at present.

Uptake of virus, following such injury, possibly involved receptors and the virus might be taken into the cell by pinocytosis. With the possible exception of TMV, facts concerning these possible routes were not known for either plant viruses or viroids. Spread through the infected plant might occur via plasma links between the cells (plasmodesmata).

Discussion then shifted to the replication of viroids. These contain no protein; no translation of RNA into protein can be detected in vitro. Their primary structure was that of a +ve strand which cannot produce protein. There was no good evidence of an instructional codon nor of stop signals. It might be that this +ve strand RNA could behave like a -ve strand, with the -ve copy acting as a template. In infected plant tissues, there was no evidence for a viroid-coded protein; one recent report of such a protein lacked force, as similar proteins could also be produced as a general pathological response following fungal or viral or double viral infections. The possibility was considered that protein did not play a role in viroid replication and that the RNA was acting by itself. There was no reverse transcriptase but an RNA-dependent replicase had been demonstrated. The possibility of a loose packaging system, using host proteins, could not be excluded.

The situation of viroids in the cell was also unclear. Infectivity was associated with chromatin. It was possible that viroids might be associated with molecules of cellular origin which could confer stability.

*Though Dr. Sanger's paper was presented at the workshop, it unfortunately does not appear in this volume.

The circularity of viroid RNA was alluded to. This circularity is associated with a high degree of secondary structure and Dr Sanger had found that up to 99.5% of viroids displayed a circular structure, unlike the position in Dr Diener's laboratory, where 80% were linear in structure. It was observed that, whereas circular structures can readily form linear structures by secondary folding, the formation of circular structures from linear structures would require a ligase. During the isolation of viroids, it appears that Mg˙ and an alkaline pH may lead to artefacts; thus 3mM Mg˙ at pH 9.0 will lead to a phosphate di-ester breakage and thus circular forms would tend to become linear.

Dr Tyrrell closed the discussion with a reference to the entirely new concepts which consideration of the viroids had induced.

DISCUSSION ON DR. CATHALA'S RECENT WORK

This began with a question on interference between scrapie and CJD; so far, this had not been demonstrated, although it was noted that TME can block scrapie replication. Dr Cathala mentioned that primary transmission of CJD to mice had been successful on two occasions in her hands, although inoculation of guinea-pigs had been unsuccessful.

Further discussion centred on the epidemiology of CJD. Australia (free from scrapie) did not have an appreciably lower prevalence of CJD than other countries; a similar lack of correlation was shown for Iceland, Japan and China (where there are no sheep but CJD is believed to be present). The incidence of CJD in vegetarians was unknown. Dr Cathala speculated that, in the elderly, patients with CJD might be misdiagnosed as Alzheimer's disease; in such a case, the patients would be consigned ot long-stay wards or to an old people's home and the final diagnosis at necropsy would never be known.

DISCUSSION ON THE DIRECTIONS OF FUTURE RESEARCH

As there had previously been much discussion on viroids, this topic
was dealt with briefly. Control of the viroid diseases of plants was
not possible as there were no effective virucides. The first aim of
viroid research would be to improve diagnosis as, at present, bio-
assay was not reliable and nucleic acid assay was not feasible in
field conditions. Screening for viroid disease would follow and, in
turn, control by exclusion of plants with viroid disease from propa-
gation. These processes would be carried out by the gardener or
agriculturist, thus demonstrating the close relationship in this area
between applied and basic research.

In scrapie, the problem of the zero phase, where no marker was
available, was important, as was the situation where experimental
animals develop severe neurological signs without corresponding neuropathology. There had been too much emphasis on the scrapie agent as
such and more work should be done on the basic biochemical defects
which resulted in death or malfunction. The nature of the molecular
lesion should be studied and neurochemical probes should be more
extensively used. Few sequential studies had been carried out and
these should be set up to study the earliest changes, particularly
those at the post-synaptic side of the synapse. The 35 nm tubular
filamentous spherical particles which have been described should be
further investigated. Dr Tyrrell suggested that the neurophysiology
of the disease should be studied. Scrapie was a steadily progressive
disease; why was this so? Does reactivation occur? As scrapie
replicates in the spleen, organ culture of that organ might be helpful.
Prolines, purine metabolism and the effect of scrapie on the enzymes
related to purine metabolism should be investigated.

In SSPE and distemper, Dr Norrby suggested that the serological
response should be studied further. Similarly, the molecular aspects
of SSPE deserved more attention and more strains of SSPE virus should
be obtained in an effort to detect markers on different strains. In vivo
changes in the virus should be studied and it was proposed that a
spectrum of SSPE virus strains should be assembled, and their relations
with the immune defence system investigated. Doubt was expressed
whether the study of tissue culture adapted virus strains was entirely

valid. The question was raised whether there were one or more strains
of wild measles virus; the answer lay in studying homogeneity of such
strains by restriction of enzyme analysis, in a similar way to that
used with herpes simplex virus.

Although doubts were expressed, it was suggested that a follow
up of normal uncomplicated measles in childhood might be attempted.
As SSPE occurs only in 1 in 1 x 10^6 children, selected cases only
could be included and attempts might be made to identify a subgroup
which might be at risk. The imperfect models of SSPE which exist in
animals should be studied further.

Dr Thiry called for a collaborative clinical/virological group
to investigate the role of oncornaviruses in man. The topics she
suggested were infertility, repeated spontaneous abortions, hydatidi-
form mole and pre-eclampsia. The investigation of newly diagnosed
patients with subacute lupus erythematosus and acute diffuse glomerulonephritis was also suggested. Protocols for sampling were needed and,
with technical improvements now available, it should be possible to
search for and assess the role of these viruses in man.

As for Maedi/Visna, it was felt that it was possible that some
disease in man might resemble this condition and this should be
studied further. In vivo variation of visna/maedi clearly needed
further attention.

As a member of the general group of topics included in this
meeting, interferon and its immunology deserved further research.

The meeting was closed by Dr Tyrrell with thanks to the partici-
pants and to the EEC.

INDEX OF SUBJECTS